Graded examples in mathematics

Fractions and Decimals

M. R. Heylings M.A., M.Sc.

Schofield & Sims Limited Huddersfield

0 7217 2323 3

First printed 1982
Reprinted 1983
Revised and reprinted 1984
Reprinted 1986
Reprinted 1987

The series **Graded examples in mathematics** comprises:

Fractions and Decimals	0 7217 2323 3
Answer Book	0 7217 2324 1
Algebra	0 7217 2325 x
Answer Book	0 7217 2326 8
Area and Volume	0 7217 2327 6
Answer Book	0 7217 2328 4
General Arithmetic	0 7217 2329 2
Answer Book	0 7217 2330 6
Geometry and Trigonometry	0 7217 2331 4
Answer Book	0 7217 2332 2
Negative Numbers and Graphs	0 7217 2333 0
Answer Book	0 7217 2334 9
Matrices and Transformations	0 7217 2335 7
Answer Book	0 7217 2336 5
Sets, Probability and Statistics	0 7217 2337 3
Answer Book	0 7217 2338 1
Revision of Topics for GCSE	0 7217 2339 x
Answer Book	0 7217 2340 3

Designed by Peter Sinclair (Design and Print) Ltd, Wetherby
Printed in England by Pindar Print Limited, Scarborough, North Yorkshire.

Author's Note

This series has been written and produced in the form of eight topic books, each offering a wealth of graded examples for pupils in the 11-16 age range; plus a further book of revision examples for those nearing examination in year 5.

There are no teaching points in the series. The intention is to meet the often heard request from teachers for a wide choice of graded examples to support their own class teaching. The contents are clearly labelled for easy use in conjunction with an existing course book; but the books can also be used as the chief source of examples, in which case the restrictions imposed by the traditional type of mathematics course book are removed and the teacher is free to organise year-by-year courses to suit the school. Used in this way, the topic-book approach offers an unusual and useful continuity of work for the class-room, for homework or for revision purposes.

The material has been tested over many years in classes ranging from mixed ability 11-year-olds to fifth formers taking public examinations. Some sections are useful for pupils of above average ability while other sections suit the needs of the less able, though it is for the middle range of ability that the series is primarily intended.

"Graded examples in mathematics" and the GCSE

The advent of GCSE has increased the demands made on the teaching and learning of mathematics. A heavy stress is laid on the content of a mathematics programme being related to applications and contexts, particularly those from everyday situations, and being such that pupils increase their self-confidence by experiencing success in their work. The exercises offered in this series are so graded as to encourage success; in addition, throughout the series, a continuing emphasis is placed on providing exercises where pupils can apply the topic being studied in a variety of contexts and situations. The series thus offers a highly flexible and motivating preparation for GCSE.

Contents

Fractions

Decimals: Preliminary Exercises

Decimals: Money and Length

Decimals: General Exercises

Decimals, Fractions and Percentages

Symbols

$=$	is equal to
$\neq$	is not equal to
$\simeq$	is approximately equal to
$<$	is less than
$\leqslant$	is less than or equal to
$\nless$	is not less than
$>$	is greater than
$\geqslant$	is greater than or equal to
$\ngtr$	is not greater than
$\Rightarrow$	implies
$\Leftarrow$	is implied by
$\rightarrow$	maps onto
$\in$	is a member of
$\notin$	is not a member of
$\subset$	is a subset of
$\not\subset$	is not a subset of
$\cap$	intersection (or overlap)
$\cup$	union
A'	the complement (or outside) of set A
$\mathscr{E}$	the Universal set
$\varnothing$ or $\{ \}$	the empty set
(x, y)	the co-ordinates of a point
$\begin{pmatrix} x \\ y \end{pmatrix}$	the components of a vector

The Greek alphabet

A	α	alpha
B	β	beta
Γ	γ	gamma
Δ	δ	delta
E	ε	epsilon
Z	ζ	zeta
H	η	eta
Θ	θ	theta
I	ι	iota
K	κ	kappa
Λ	λ	lambda
M	μ	mu
N	ν	nu
Ξ	ξ	xi
O	o	omicron
Π	π	pi
P	ρ	rho
Σ	σ, ς	sigma
T	τ	tau
Υ	υ	upsilon
Φ	ϕ, φ	phi
X	χ	chi
Ψ	ψ	psi
Ω	ω	omega

Fractions

Shaded fractions
Whole numbers and fractions
Equivalent fractions
Multiplying and dividing fractions
Halves, quarters, eighths etc.
Addition of fractions
Subtraction of fractions
Fractions of quantities

Shaded fractions

Write the fraction of each shape which is **a** *shaded*
b *not shaded.*

1

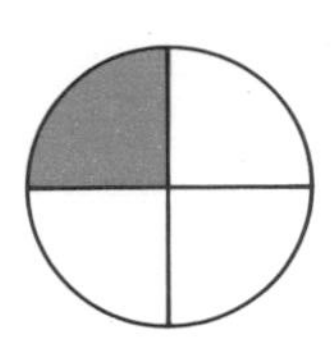

2

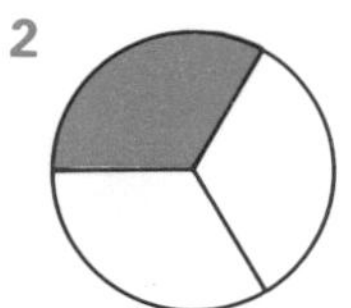

3

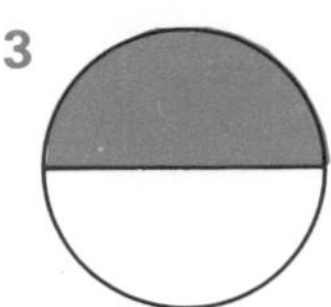

4

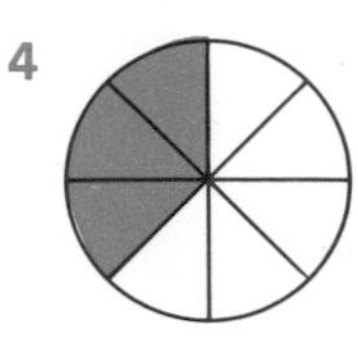

5

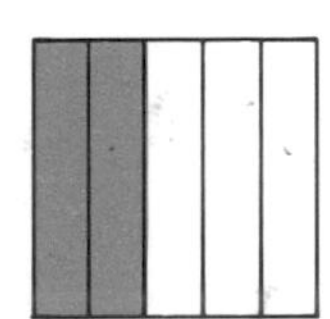

6

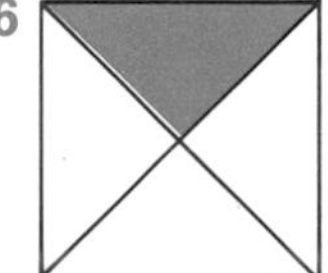

7

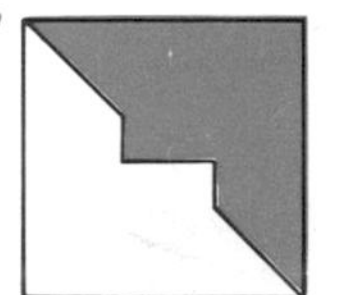

8

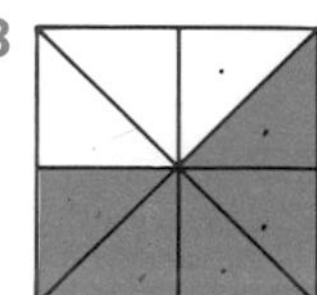

9

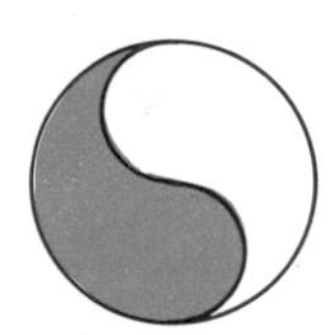

10

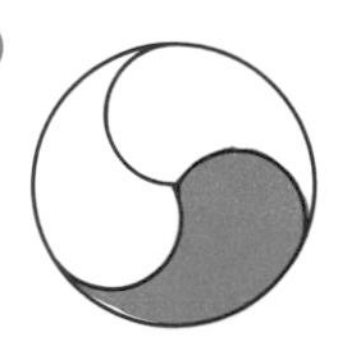

11

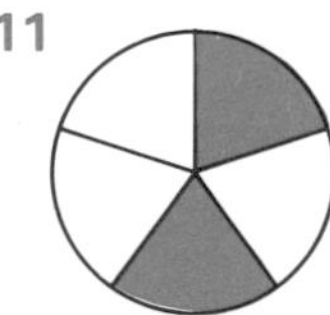

12

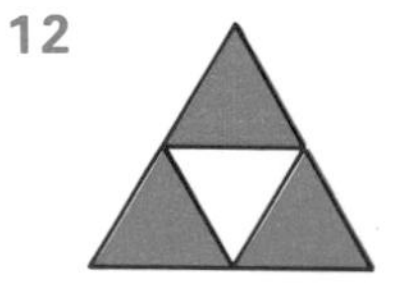

13

14

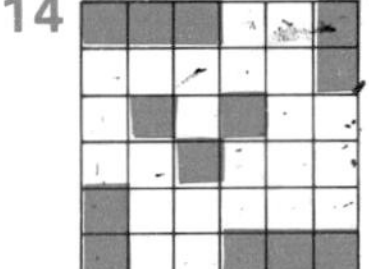

15

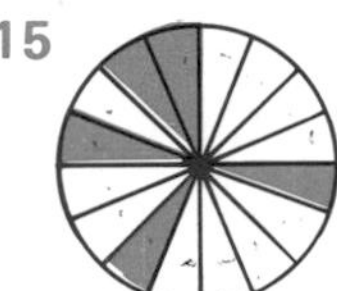

16

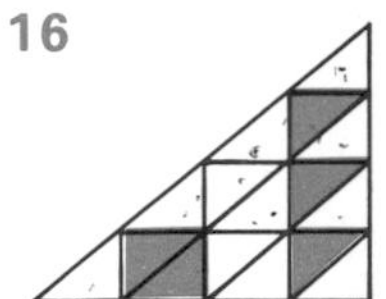

For these next shapes, write the fraction which is **a** *blank*
b *shaded*
c *dotted.*

17

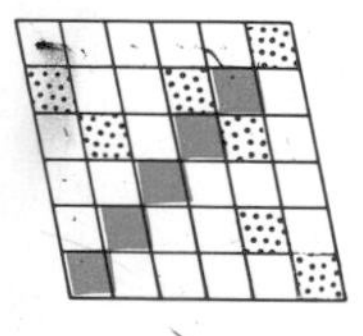

18

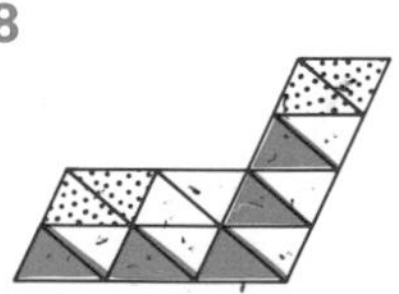

19

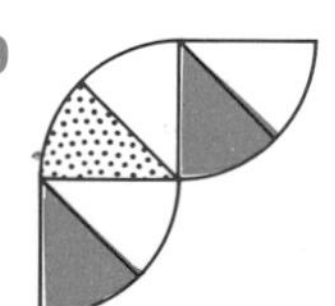

20

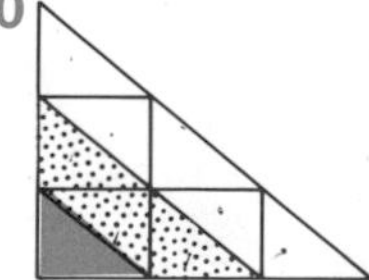

21

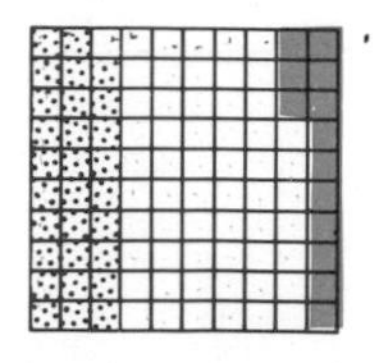

22

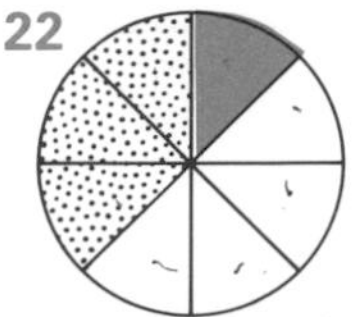

23

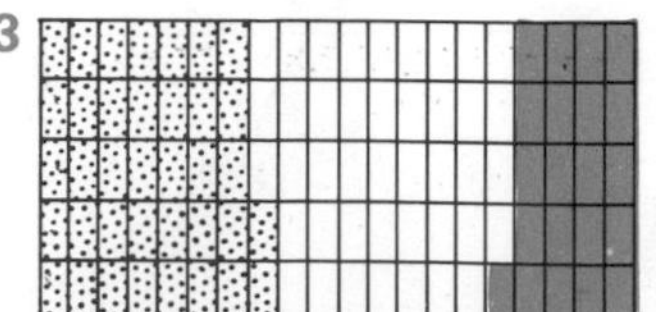

24

Shaded fractions

25 Each line is divided into equal sections.
What fraction is coloured?

a 

b 

c

d 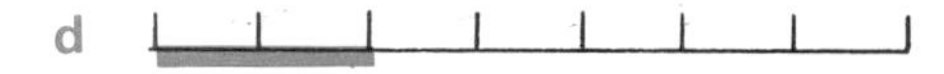

e

f 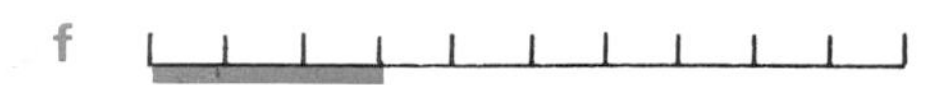

g

h

i

j

Copy these shapes and shade in the fractions given.

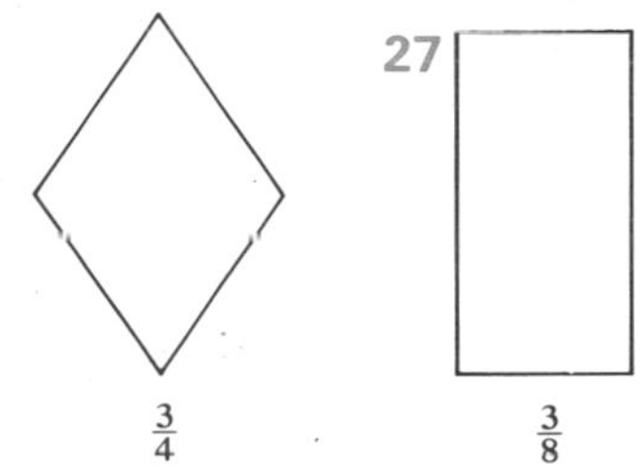

26 $\frac{3}{4}$

27 $\frac{3}{8}$

28 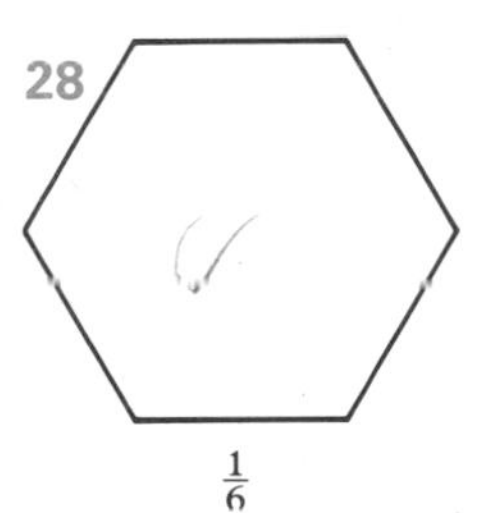

$\frac{1}{6}$

29 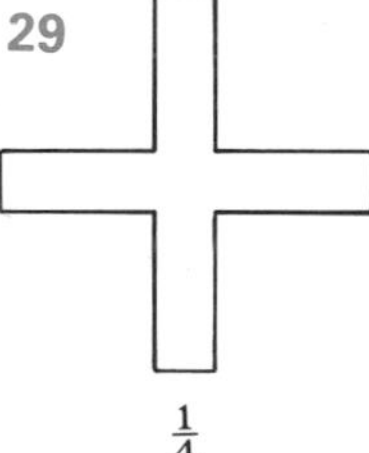

$\frac{1}{4}$

30 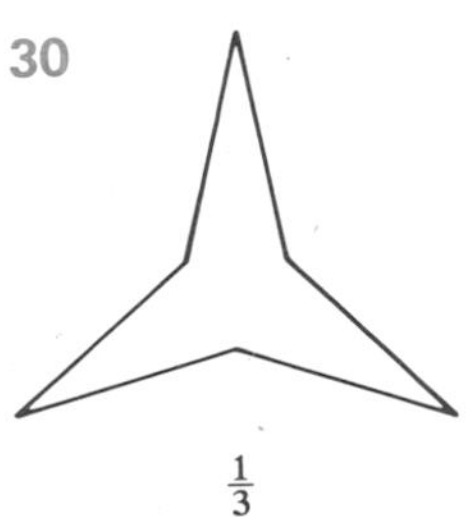

$\frac{1}{3}$

31 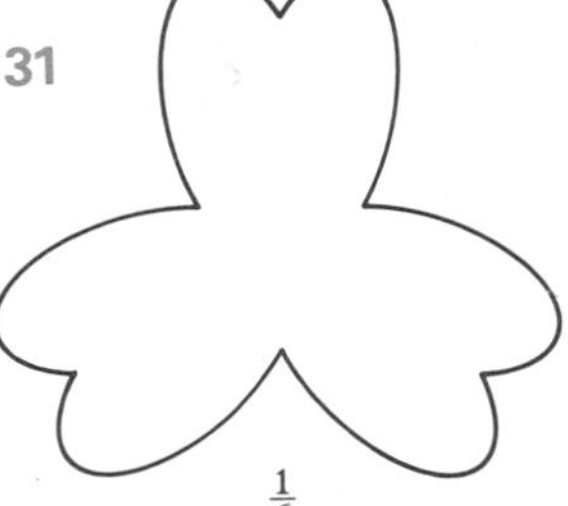

$\frac{1}{6}$

32 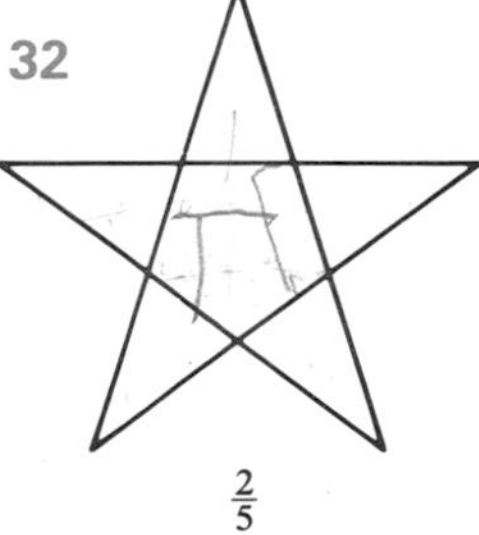

$\frac{2}{5}$

33 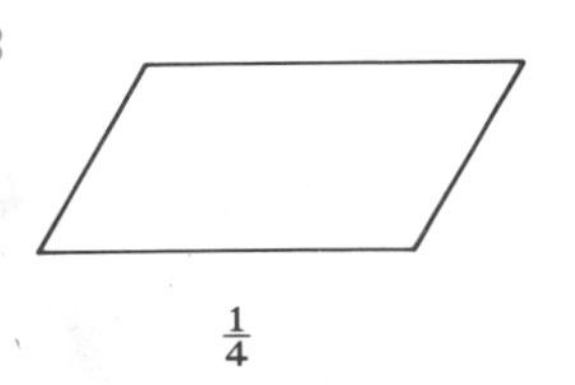

$\frac{1}{4}$

34 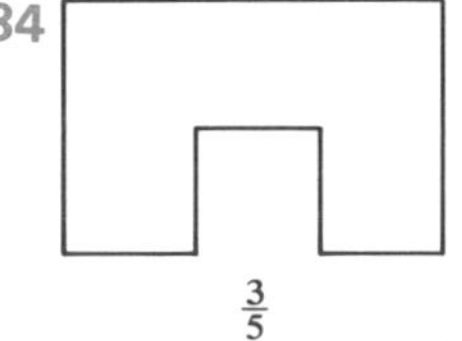

$\frac{3}{5}$

35 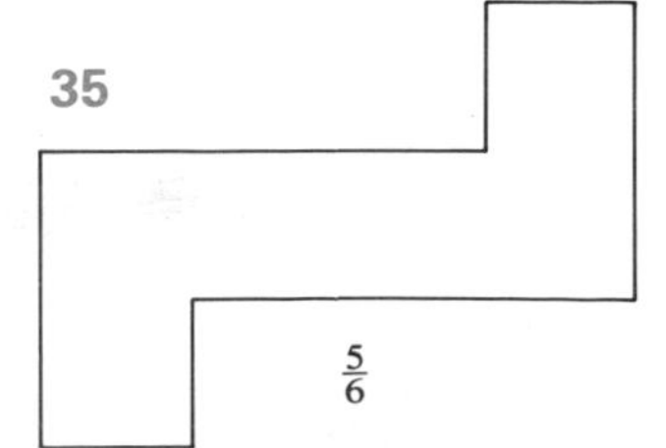

$\frac{5}{6}$

Shaded fractions

36 What fraction of the word MISSISSIPPI is made up of Ss?

37 What fraction of the word ABRACADABRA is made up of As?

38 My mother buys 16 pieces of fruit, of which 7 are apples, 5 are oranges and 4 are bananas. What fraction of the fruit is made up of oranges?

39 My telephone number is 0484-83-5643. What fraction of this number is made up of 4s?

40 Jane spent £12 on Saturday, of which £2 was spent on books, £7 on clothes and £3 on records. What fraction of her money was spent on clothes?

41 A man has six spanners, two screw-drivers, a hammer, two saws, a drill and a wrench. What fraction of his total tool kit is spanners?

42 A farmer has 20 cows, 35 sheep, 12 pigs and 2 goats. What fraction of his animals is sheep?

43 The time Simon spent on his homework last week was as follows:
2h on science $1\frac{1}{2}$h on English $1\frac{1}{2}$h on geography
2h on maths 1h on French 1h on history.
What fraction of his homework time was spent on maths?

44 At a youth group 12 members drink lemonade, 8 drink orange squash, 5 drink ginger-ale and 2 drink limeade. What fraction of the members drink orange squash?

45 Mr Dickenson died last week after spending 31 years working in a coal-mine. In addition he worked for 9 years in a factory and for 6 years as a gardener. He was retired for 12 years and got his first job when he was 16 years old.
a How long did Mr Dickenson live?
b What fraction of his life was spent as a miner?

46 Mrs Phillips spent 45p on a loaf of bread, 35p on milk, 65p on butter, 48p on sugar, 87p on breakfast cereals, 37p on tea, 62p on rice and 55p on flour.
a How much in pence did she spend?
b What fraction of her total was spent on tea?

47 Of the 365 days of last year, my car was used on 214 days to travel to work and on 88 days for pleasure and holidays. It was at the garage for repair on 12 days, and on all other days it was not used.
a On how many days did I not use my car?
b For what fraction of the year was my car not used?

48 Here is how John Townsend spent his day yesterday.
7h at school 8h sleeping $1\frac{1}{2}$h homework $2\frac{1}{2}$h eating
The rest of the day was spent on leisure activities.
What fraction of the day did he spend in this way?

Whole numbers and fractions

Part 1

1 Use the diagrams to help you answer the following questions.

a How many halves are there in $2\frac{1}{2}$?

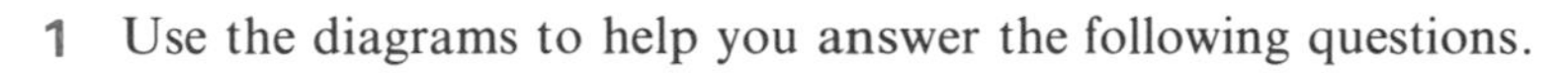

b How many quarters are there in $1\frac{3}{4}$?

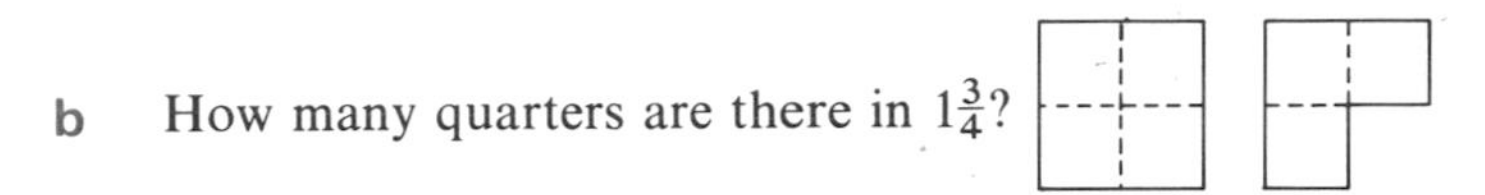

c How many thirds are there in $2\frac{2}{3}$?

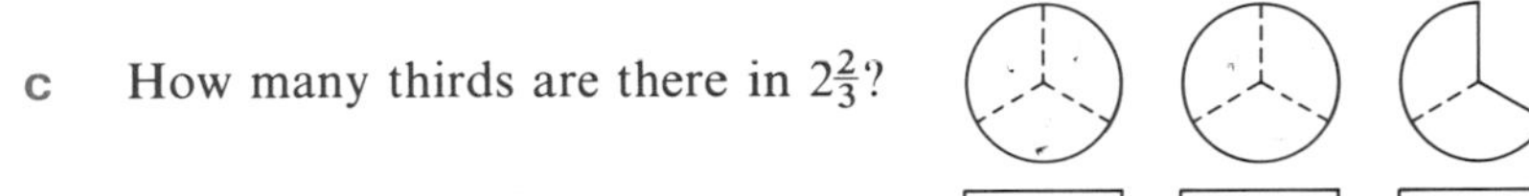

d How many halves are there in 3?

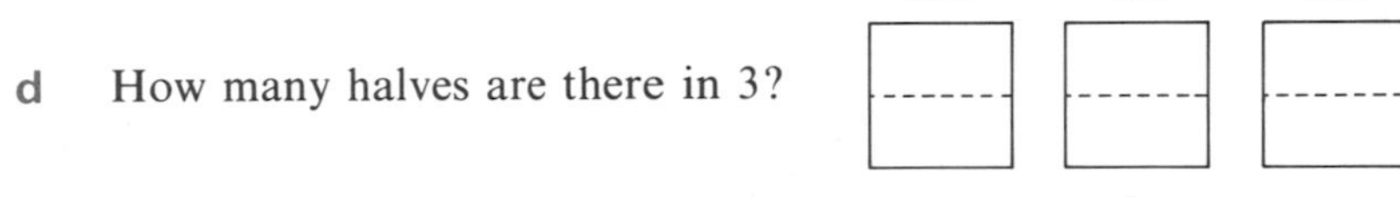

e How many fifths are there in $1\frac{2}{5}$?

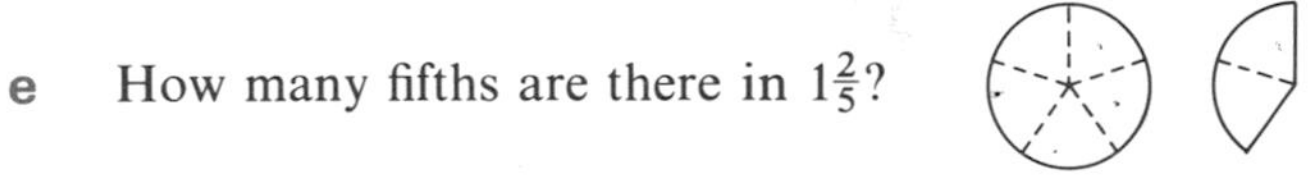

f How many sixths are there in 4?

g How many quarters are there in 2?

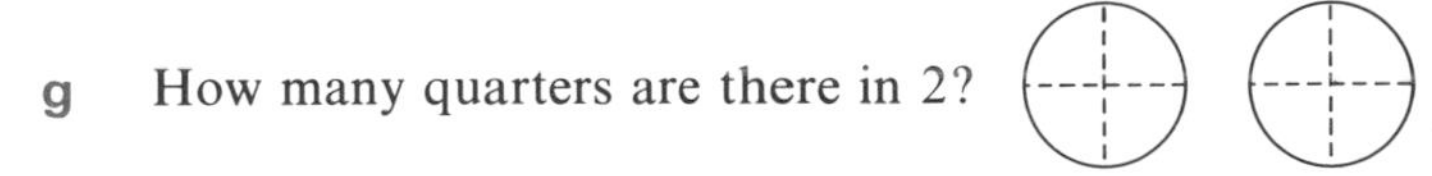

h How many thirds are there in $3\frac{1}{3}$?

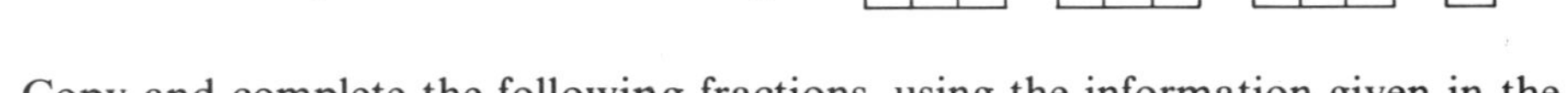

2 Copy and complete the following fractions, using the information given in the diagrams.

a $3\frac{1}{4} = \frac{}{4}$

b $1\frac{2}{3} = \frac{}{3}$

c $2\frac{3}{5} = \frac{}{5}$

d $1\frac{3}{8} = \frac{}{8}$

e $3\frac{1}{5} = \frac{}{5}$

f $1\frac{5}{6} = \frac{}{6}$

g $3\frac{7}{8} = \frac{}{8}$

h $2\frac{3}{4} = \frac{}{4}$

i $3\frac{7}{10} = \frac{}{10}$

j $1\frac{9}{10} = \frac{}{10}$

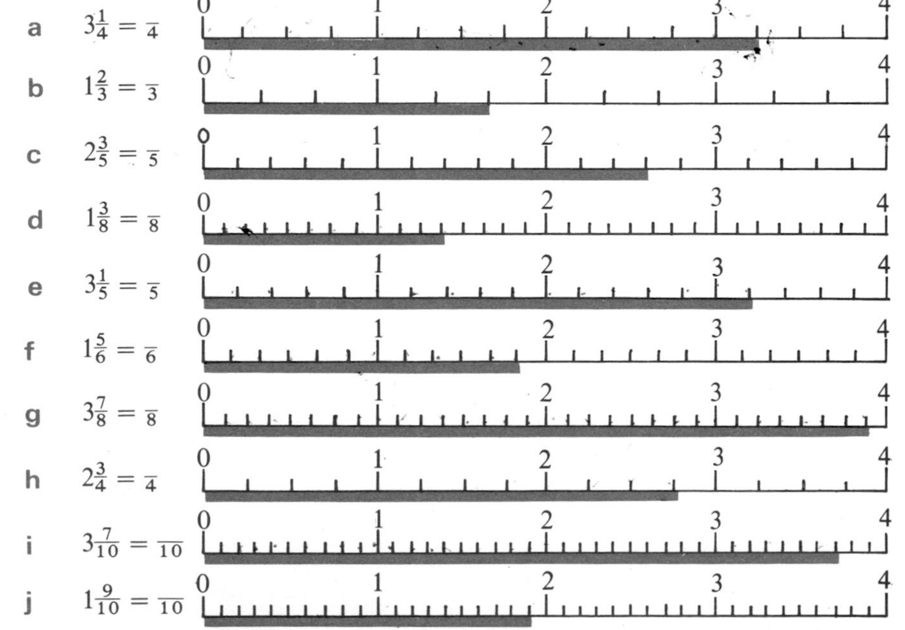

Whole numbers and fractions

3 How many halves are there in:

a 2 b 5 c $5\frac{1}{2}$ d 6 e $6\frac{1}{2}$

f 10 g $10\frac{1}{2}$ h $3\frac{1}{2}$ i $8\frac{1}{2}$ j $4\frac{1}{2}$?

4 How many thirds are there in:

a 2 b $2\frac{1}{3}$ c 3 d $3\frac{2}{3}$ e 6

f $6\frac{2}{3}$ g 4 h $4\frac{1}{3}$ i 10 j $10\frac{2}{3}$?

5 How many quarters are there in:

a 2 b $2\frac{1}{4}$ c $2\frac{3}{4}$ d 5 e $5\frac{3}{4}$

f 3 g $4\frac{1}{4}$ h $7\frac{3}{4}$ i $8\frac{1}{4}$ j $6\frac{3}{4}$?

6 How many fifths are there in:

a 2 b $2\frac{2}{5}$ c $2\frac{4}{5}$ d 3 e $3\frac{3}{5}$

f $5\frac{1}{5}$ g $6\frac{3}{5}$ h 7 i $4\frac{2}{5}$ j $5\frac{4}{5}$?

7 How many tenths are there in:

a 2 b $2\frac{3}{10}$ c 2.3 d $4\frac{7}{10}$ e 4.7

f $3\frac{9}{10}$ g 3.9 h 5.4 i $1\frac{3}{10}$ j $6\frac{7}{10}$?

8 Copy and complete, changing the mixed numbers into *top-heavy* fractions.

a $3\frac{1}{2} = \frac{}{2}$ b $2\frac{2}{3} = \frac{}{3}$ c $4\frac{2}{5} = \frac{}{5}$ d $3\frac{7}{10} = \frac{}{10}$

e $7\frac{3}{4} = \frac{}{4}$ f $1\frac{5}{6} = \frac{}{6}$ g $9\frac{2}{3} = \frac{}{3}$ h $4\frac{4}{5} = \frac{}{5}$

i $5\frac{3}{10} = \frac{}{10}$ j $7\frac{2}{3} = \frac{}{3}$ k $4\frac{3}{4} = \frac{}{4}$ l $1\frac{5}{9} = \frac{}{9}$

m $12\frac{3}{4} = \frac{}{4}$ n $3.6 = \frac{}{10}$ o $3\frac{4}{9} = \frac{}{9}$ p $8.6 = \frac{}{10}$

9 How many millimetres are there in:

a 2 cm b $2\frac{3}{10}$ cm c 2.3 cm d $4\frac{7}{10}$ cm e 4.7 cm

f $5\frac{9}{10}$ cm g $1\frac{3}{10}$ cm h $8\frac{7}{10}$ cm i 3.6 cm j $1\frac{9}{10}$ cm?

10 How many months are there in:

a $1\frac{5}{12}$ years b $2\frac{1}{12}$ years c $2\frac{7}{12}$ years d $3\frac{11}{12}$ years e $5\frac{1}{12}$ years

f $4\frac{5}{12}$ years g $1\frac{11}{12}$ years h $3\frac{5}{12}$ years i $6\frac{1}{12}$ years j $4\frac{11}{12}$ years?

11 How many days are there in:

a 2 weeks b $2\frac{3}{7}$ weeks c 3 weeks d $3\frac{4}{7}$ weeks e $5\frac{2}{7}$ weeks

f $1\frac{6}{7}$ weeks g $7\frac{1}{7}$ weeks h $4\frac{6}{7}$ weeks i $9\frac{5}{7}$ weeks j $10\frac{2}{7}$ weeks?

12 Copy and complete, changing the mixed numbers into *top-heavy* fractions.

a $5\frac{6}{7} = \frac{}{7}$ b $8\frac{4}{9} = \frac{}{9}$ c $5\frac{3}{5} = \frac{}{5}$ d $9\frac{5}{8} = \frac{}{8}$

e $11\frac{1}{9} = \frac{}{9}$ f $6\frac{6}{7} = \frac{}{7}$ g $3\frac{3}{8} = \frac{}{8}$ h $6\frac{5}{12} = \frac{}{12}$

i $8\frac{8}{15} = \frac{}{15}$ j $6\frac{11}{20} = \frac{}{20}$ k $7\frac{19}{25} = \frac{}{25}$ l $3\frac{11}{15} = \frac{}{15}$

m $11\frac{5}{12} = \frac{}{12}$ n $4\frac{9}{14} = \frac{}{14}$ o $6\frac{11}{19} = \frac{}{19}$ p $12\frac{13}{15} = \frac{}{15}$

Whole numbers and fractions

Part 2

1 Use the diagrams to help you change these *top-heavy* fractions into mixed numbers.

a $\frac{5}{4} =$ 

b $\frac{7}{2} =$ 

c $\frac{8}{3} =$

d $\frac{9}{5} =$ 

e $\frac{11}{4} =$ 

f $\frac{6}{3} =$

g $\frac{8}{2} =$  

h $\frac{7}{5} =$

i $\frac{13}{8} =$ 

j $\frac{19}{10} =$

k $\frac{17}{7} =$

l $\frac{13}{6} =$

m $\frac{13}{4} =$

n $\frac{13}{5} =$

2 Write these fractions as whole numbers.

a	six halves	b	eight quarters	c	six thirds
d	forty tenths	e	twenty fifths	f	forty eighths
g	twenty quarters	h	thirty thirds	i	sixty twelfths
j	eighteen thirds	k	forty quarters	l	thirty fifths

3 Write these fractions as mixed numbers.

a	seven halves	b	ten thirds	c	nine quarters
d	twelve fifths	e	nine eighths	f	twenty ninths
g	twenty sevenths	h	fifteen quarters	i	seventeen halves

j $\frac{7}{2}$ k $\frac{10}{3}$ l $\frac{9}{4}$ m $\frac{12}{5}$ n $\frac{9}{8}$

o $\frac{20}{9}$ p $\frac{25}{2}$ q $\frac{32}{3}$ r $\frac{47}{5}$ s $\frac{31}{7}$

t $\frac{41}{6}$ u $\frac{77}{10}$ v $\frac{149}{12}$ w $\frac{79}{5}$ x $\frac{147}{8}$

4 Are the following statements *true* or *false*?

a $\frac{9}{2} = 4\frac{1}{2}$ b $\frac{7}{3} = 2\frac{1}{3}$ c $\frac{9}{4} = 2\frac{1}{2}$ d $\frac{23}{5} = 4\frac{1}{3}$

e $\frac{38}{7} = 5\frac{3}{7}$ f $\frac{11}{4} = 5\frac{1}{2}$ g $\frac{49}{8} = 6\frac{1}{2}$ h $\frac{25}{2} = 12\frac{1}{2}$

i $\frac{61}{8} = 7\frac{1}{2}$ j $\frac{73}{10} = 7.3$ k $\frac{82}{9} = 9\frac{1}{3}$ l $\frac{100}{3} = 30\frac{1}{3}$

m $\frac{63}{12} = 5\frac{3}{4}$ n $\frac{46}{10} = 4.6$ o $\frac{91}{8} = 11\frac{3}{8}$ p $\frac{27}{4} = 6\frac{3}{4}$

q $\frac{57}{10} = 5.7$ r $\frac{81}{2} = 4\frac{1}{2}$ s $\frac{91}{9} = 1\frac{1}{9}$ t $\frac{24}{10} = 0.24$

u $\frac{101}{12} = 8\frac{7}{12}$ v $\frac{99}{5} = 20\frac{4}{5}$ w $\frac{77}{20} = 4\frac{17}{20}$ x $\frac{121}{3} = 40\frac{1}{3}$

y $\frac{86}{15} = 5\frac{9}{15}$ z $\frac{45}{14} = 3\frac{1}{14}$

Equivalent fractions

Part 1

1 Copy each pair of equivalent fractions and fill in the missing number.
Use the diagrams to help you.

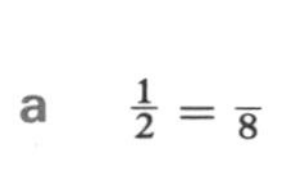

a $\frac{1}{2} = \frac{}{8}$

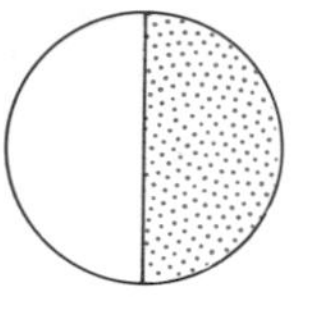
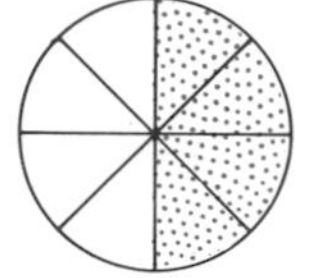

b $\frac{1}{4} = \frac{}{8}$

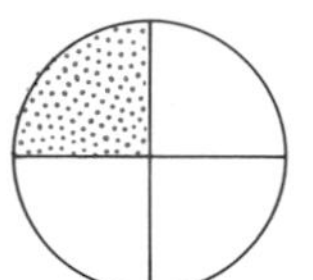
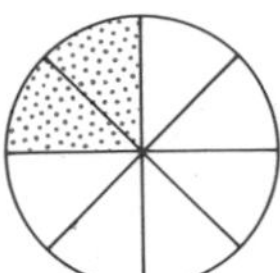

c $\frac{3}{4} = \frac{}{8}$

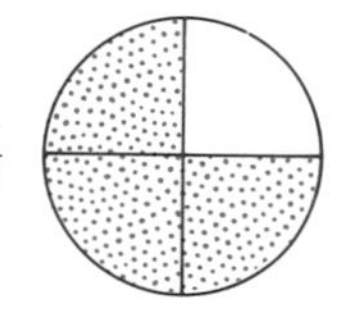
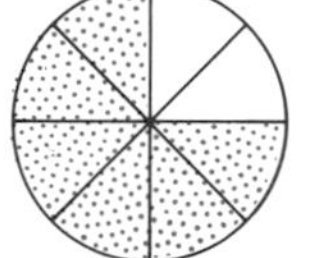

d $\frac{3}{8} = \frac{}{16}$

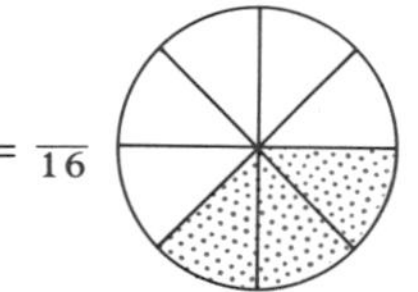
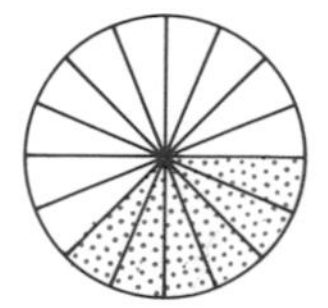

e $\frac{1}{3} = \frac{}{6}$

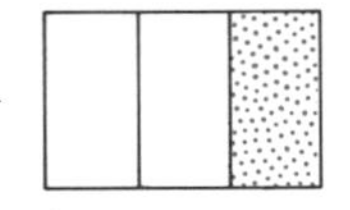

f $\frac{2}{3} = \frac{}{6}$

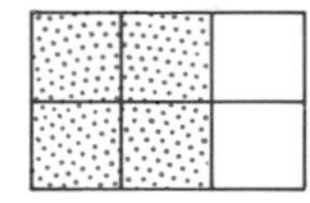

g $\frac{1}{6} = \frac{}{12}$

h $\frac{5}{6} = \frac{}{12}$

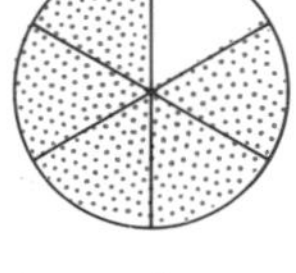
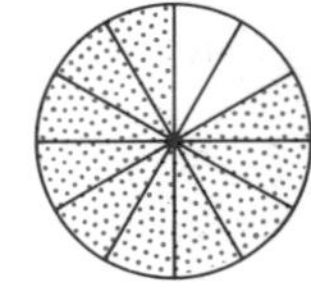

i $\frac{3}{4} = \frac{}{12}$

j $\frac{1}{4} = \frac{}{12}$

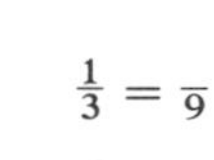

k $\frac{1}{3} = \frac{}{9}$

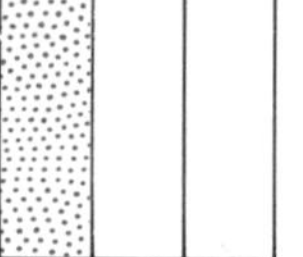

l $\frac{2}{3} = \frac{}{9}$

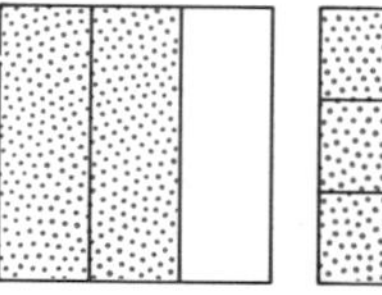

m $\frac{2}{5} = \frac{}{10}$

n $\frac{3}{5} = \frac{}{10}$

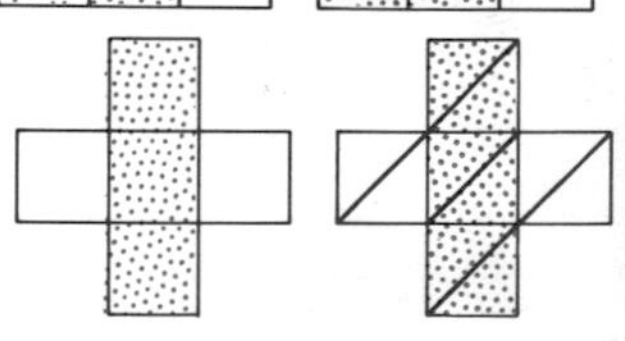

o $\frac{1}{2} = \frac{}{12}$

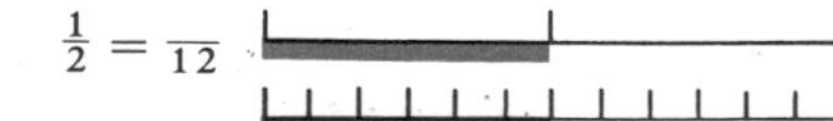

p $\frac{2}{3} = \frac{}{12}$

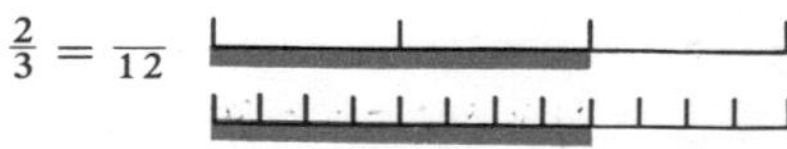

q $\frac{1}{5} = \frac{}{10}$

r $\frac{3}{5} = \frac{}{15}$

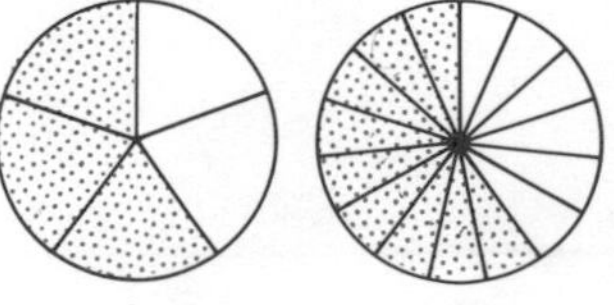

s $\frac{1}{4} = \frac{}{16}$

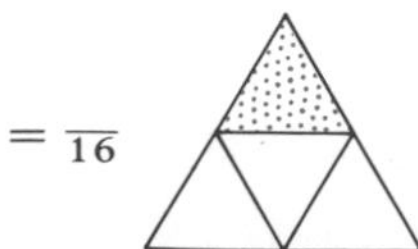

t $\frac{3}{4} = \frac{}{16}$

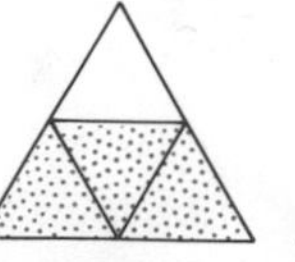
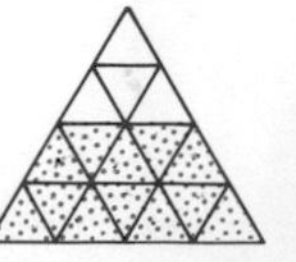

Equivalent fractions

Copy this table. Label it **fractions shaded**.

Complete a row for each group of shapes, giving the fraction of each shape which is *shaded*.

fractions shaded	a	b	c	d
Shape **2**	=	=	=	
Shape **3**	=	=	=	
Shape **4**	=	=	=	
Shape **5**	=	=	=	

2 a 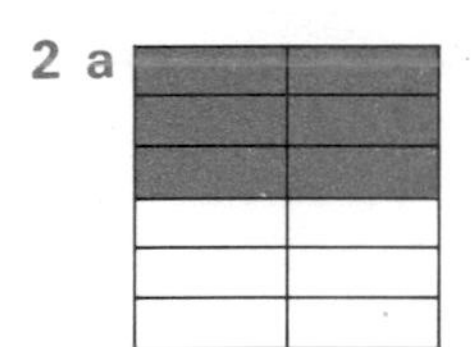 b 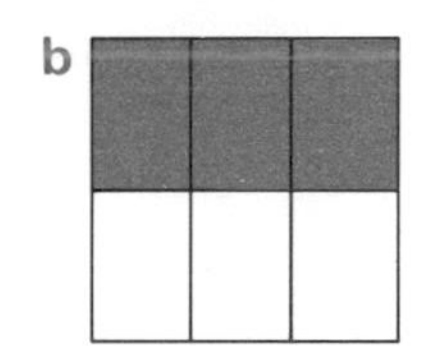 c 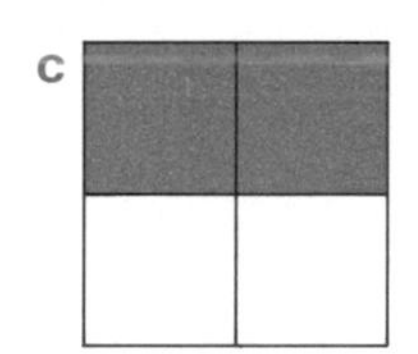 d 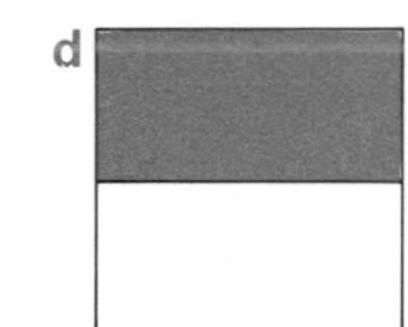

3 a 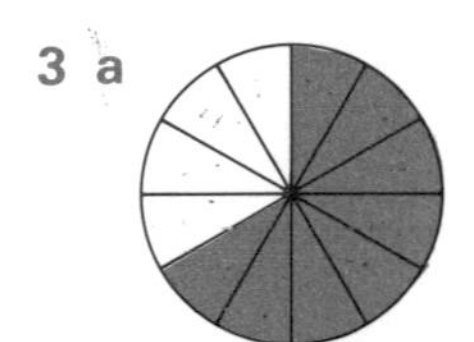 b 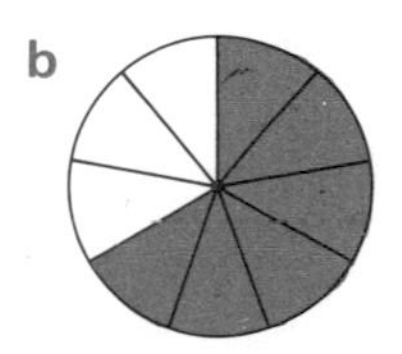 c 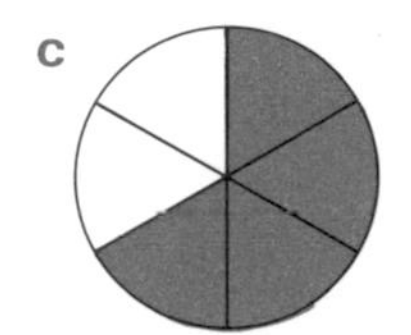 d 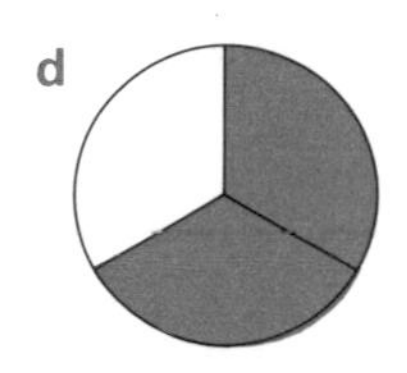

4 a 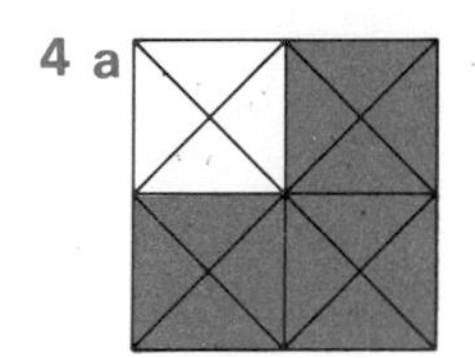 b 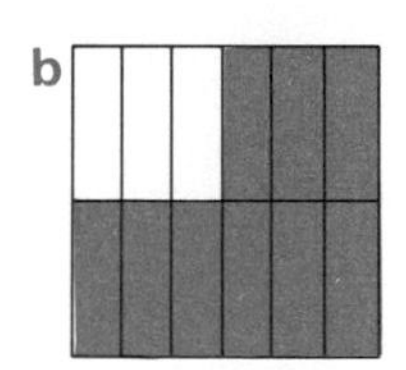 c 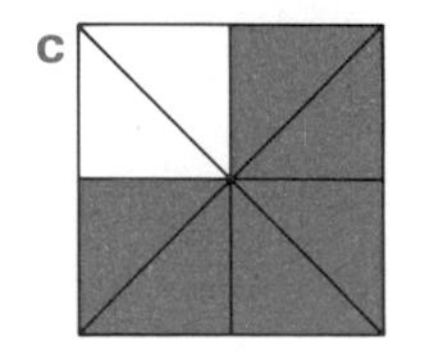 d 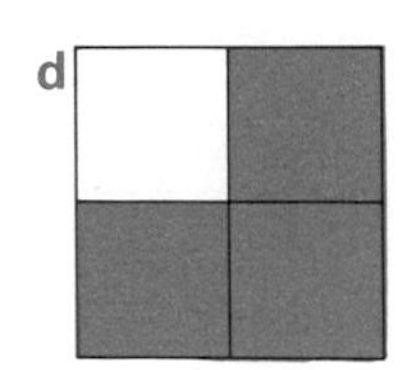

5 a 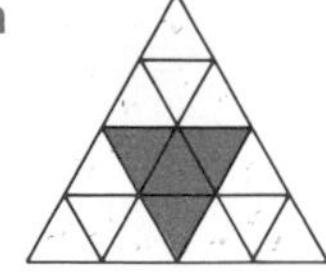 b c d

6 Copy the above table again, but now label it **fractions unshaded**.
Complete the rows by looking at the *unshaded* parts of the drawings.

Equivalent fractions

7 This diagram is called a **fraction chart**.

ONE WHOLE

$\frac{1}{2}$

$\frac{1}{3}$

$\frac{1}{4}$

$\frac{1}{5}$

$\frac{1}{6}$

$\frac{1}{7}$

$\frac{1}{8}$

$\frac{1}{9}$

$\frac{1}{10}$

$\frac{1}{11}$

$\frac{1}{12}$

$\frac{1}{13}$

$\frac{1}{14}$

$\frac{1}{15}$

$\frac{1}{16}$

Copy and complete these sets of equivalent fractions using the fraction chart.
A ruler placed down the chart will make it easier for you.

a $\frac{1}{2} = \frac{}{4} = \frac{}{6} = \frac{}{8} = \frac{}{10} = \frac{}{12} = \frac{}{14} = \frac{}{16}$

b $\frac{1}{3} = \frac{}{6} = \frac{}{9} = \frac{}{12} = \frac{}{15}$

c $\frac{1}{4} = \frac{}{8} = \frac{}{12} = \frac{}{16}$

d $\frac{1}{5} = \frac{}{10} = \frac{}{15}$

e $\frac{1}{6} = \frac{\quad}{12}$ f $\frac{1}{7} = \frac{\quad}{14}$ g $\frac{1}{8} = \frac{\quad}{16}$

h $\frac{3}{4} = \frac{\quad}{8} = \frac{\quad}{12} = \frac{\quad}{16}$ i $\frac{2}{3} = \frac{\quad}{6} = \frac{\quad}{9} = \frac{\quad}{12} = \frac{\quad}{15}$ j $\frac{4}{7} = \frac{\quad}{14}$

k $\frac{3}{5} = \frac{\quad}{10} = \frac{\quad}{15}$ l $\frac{3}{8} = \frac{\quad}{16}$ m $\frac{2}{5} = \frac{\quad}{10} = \frac{\quad}{15}$

n $\frac{5}{6} = \frac{\quad}{12}$ o $\frac{2}{7} = \frac{\quad}{14}$ p $\frac{4}{5} = \frac{\quad}{10} = \frac{\quad}{15}$

q $\frac{5}{7} = \frac{\quad}{14}$ r $\frac{5}{8} = \frac{\quad}{16}$ s $\frac{7}{8} = \frac{\quad}{16}$

8 Use the fraction chart to say whether the following statements are *true* or *false*.

a $\frac{5}{6} = \frac{10}{12}$ b $\frac{4}{7} = \frac{8}{14}$ c $\frac{3}{5} = \frac{7}{15}$ d $\frac{2}{3} = \frac{3}{10}$

e $\frac{2}{5} = \frac{4}{10}$ f $\frac{2}{3} = \frac{8}{12}$ g $\frac{3}{4} = \frac{11}{13}$ h $\frac{4}{9} = \frac{7}{15}$

i $\frac{3}{7} = \frac{6}{14}$ j $\frac{4}{5} = \frac{7}{12}$ k $\frac{3}{8} = \frac{7}{12}$ l $\frac{3}{8} = \frac{6}{16}$

m $\frac{1}{2} = \frac{6}{11}$ n $\frac{4}{7} = \frac{9}{14}$ o $\frac{3}{12} = \frac{4}{16}$ p $\frac{6}{8} = \frac{9}{12}$

9 The sign > means 'is bigger than'. The sign < means 'is smaller than'.
Use the fraction chart to say whether the following are *true* or *false*.

a $\frac{1}{2} < \frac{4}{7}$ b $\frac{3}{4} > \frac{7}{11}$ c $\frac{2}{7} > \frac{4}{15}$ d $\frac{6}{10} < \frac{8}{15}$

e $\frac{4}{6} < \frac{8}{11}$ f $\frac{7}{11} > \frac{8}{12}$ g $\frac{2}{16} > \frac{1}{9}$ h $\frac{7}{15} < \frac{3}{7}$

i $\frac{1}{2} < \frac{2}{3}$ j $\frac{2}{3} < \frac{3}{4}$ k $\frac{2}{5} > \frac{3}{8}$ l $\frac{1}{4} < \frac{2}{8}$

m $\frac{5}{6} > \frac{9}{12}$ n $\frac{3}{8} < \frac{6}{16}$ o $\frac{3}{4} < \frac{13}{16}$ p $\frac{2}{7} > \frac{3}{10}$

10 Insert the sign =, > or < between the pairs of fractions in order to make the statement *true*.
Use the fraction chart to help you.

a $\frac{3}{16}$ $\frac{2}{11}$ b $\frac{8}{15}$ $\frac{6}{11}$ c $\frac{2}{11}$ $\frac{1}{5}$ d $\frac{3}{4}$ $\frac{12}{16}$

e $\frac{1}{4}$ $\frac{3}{11}$ f $\frac{2}{9}$ $\frac{3}{13}$ g $\frac{3}{5}$ $\frac{9}{15}$ h $\frac{1}{7}$ $\frac{2}{13}$

i $\frac{2}{3}$ $\frac{8}{12}$ j $\frac{8}{9}$ $\frac{14}{15}$ k $\frac{2}{3}$ $\frac{6}{9}$ l $\frac{5}{6}$ $\frac{11}{12}$

m $\frac{1}{2}$ $\frac{3}{7}$ n $\frac{5}{8}$ $\frac{14}{16}$ o $\frac{4}{5}$ $\frac{11}{15}$ p $\frac{3}{4}$ $\frac{10}{13}$

Part 2 Cancelling

1 Copy and complete each pair of fractions.

a $\frac{9}{12} = \frac{\quad}{4}$ b $\frac{15}{20} = \frac{\quad}{4}$ c $\frac{12}{15} = \frac{\quad}{5}$ d $\frac{6}{9} = \frac{\quad}{3}$

e $\frac{16}{24} = \frac{\quad}{3}$ f $\frac{25}{30} = \frac{\quad}{6}$ g $\frac{12}{18} = \frac{\quad}{3}$ h $\frac{15}{24} = \frac{\quad}{8}$

i $\frac{21}{27} = \frac{\quad}{9}$ j $\frac{14}{24} = \frac{\quad}{12}$ k $\frac{3}{12} = \frac{\quad}{4}$ l $\frac{5}{15} = \frac{\quad}{3}$

m $\frac{8}{12} = \frac{\quad}{3}$ n $\frac{8}{10} = \frac{\quad}{5}$ o $\frac{18}{24} = \frac{\quad}{4}$ p $\frac{20}{25} = \frac{\quad}{5}$

q $\frac{35}{42} = \frac{\quad}{6}$ r $\frac{40}{64} = \frac{\quad}{8}$ s $\frac{60}{84} = \frac{\quad}{7}$ t $\frac{48}{108} = \frac{\quad}{9}$

u $\frac{45}{48} = \frac{\quad}{16}$ v $\frac{39}{52} = \frac{\quad}{4}$ w $\frac{48}{64} = \frac{\quad}{4}$ x $\frac{45}{144} = \frac{\quad}{16}$

y $\frac{48}{72} = \frac{\quad}{3}$ z $\frac{76}{92} = \frac{\quad}{23}$

Equivalent fractions

2 Find equivalent fractions by cancelling as far as possible.

a $\frac{6}{8}$ b $\frac{9}{12}$ c $\frac{2}{14}$ d $\frac{3}{9}$ e $\frac{10}{15}$
f $\frac{11}{22}$ g $\frac{15}{20}$ h $\frac{7}{35}$ i $\frac{14}{21}$ j $\frac{15}{25}$
k $\frac{16}{24}$ l $\frac{15}{45}$ m $\frac{12}{32}$ n $\frac{18}{63}$ o $\frac{80}{90}$
p $\frac{12}{24}$ q $\frac{30}{36}$ r $\frac{48}{60}$ s $\frac{24}{54}$ t $\frac{56}{77}$
u $\frac{56}{84}$ v $\frac{84}{112}$ w $\frac{99}{108}$ x $\frac{77}{121}$ y $\frac{72}{90}$
z $\frac{45}{105}$

3 Cancel these fractions as far as possible.

a $\frac{18}{24}$ b $\frac{20}{35}$ c $\frac{40}{48}$ d $\frac{35}{49}$ e $\frac{27}{45}$
f $\frac{100}{160}$ g $\frac{140}{200}$ h $\frac{75}{150}$ i $\frac{54}{72}$ j $\frac{48}{80}$
k $\frac{36}{90}$ l $\frac{75}{200}$ m $\frac{48}{120}$ n $\frac{27}{108}$ o $\frac{90}{126}$
p $\frac{42}{105}$ q $\frac{60}{156}$ r $\frac{96}{216}$ s $\frac{168}{480}$ t $\frac{105}{189}$
u $\frac{192}{448}$ v $\frac{252}{308}$ w $\frac{441}{567}$ x $\frac{168}{1008}$ y $\frac{1015}{1827}$
z $\frac{315}{1980}$

4 Change these percentages to fractions and cancel as far as possible.

a 50% b 30% c 70% d 20% e 80%
f 15% g 45% h 75% i 55% j 85%
k 25% l 16% m 24% n 36% o 48%
p 72% q 96% r 64% s $1\frac{1}{2}$% t $3\frac{1}{2}$%
u $12\frac{1}{2}$% v $37\frac{1}{2}$% w $62\frac{1}{2}$% x $1\frac{1}{3}$% y $33\frac{1}{3}$%
z $66\frac{2}{3}$%

5 Change these fractions to percentages.

a $\frac{27}{100} =$ % b $\frac{51}{100} =$ %
c $\frac{89}{100} =$ % d $\frac{9}{10} = \frac{\quad}{100} =$ %
e $\frac{7}{10} = \frac{\quad}{100} =$ % f $\frac{1}{10} = \frac{\quad}{100} =$ %
g $\frac{2}{5} = \frac{\quad}{100} =$ % h $\frac{3}{5} = \frac{\quad}{100} =$ %
i $\frac{4}{5} = \frac{\quad}{100} =$ % j $\frac{7}{20} = \frac{\quad}{100} =$ %
k $\frac{11}{20} = \frac{\quad}{100} =$ % l $\frac{13}{20} = \frac{\quad}{100} =$ %
m $\frac{3}{25} = \frac{\quad}{100} =$ % n $\frac{7}{25} = \frac{\quad}{100} =$ %
o $\frac{21}{25} = \frac{\quad}{100} =$ % p $\frac{8}{200} = \frac{\quad}{100} =$ %
q $\frac{7}{200} = \frac{\quad}{100} =$ % r $\frac{25}{200} = \frac{\quad}{100} =$ %
s $\frac{24}{300} = \frac{\quad}{100} =$ % t $\frac{15}{300} = \frac{\quad}{100} =$ %
u $\frac{22}{400} = \frac{\quad}{100} =$ % v $\frac{1}{4} = \frac{\quad}{100} =$ %
w $\frac{1}{8} = \frac{\quad}{100} =$ % x $\frac{3}{8} = \frac{\quad}{100} =$ %

Equivalent fractions

6 Express each of your answers in its lowest terms.

a Mrs Wardle spent £24 on food, of which £4 was for meat.
What fraction did she spend (i) on meat (ii) on other types of food?

b A man travels 120 miles to Glasgow, of which 80 miles is by motorway.
What fraction is (i) motorway (ii) not motorway?

c A school accepts 180 new pupils of which 75 are boys.
What fraction of these pupils are (i) boys (ii) girls?

d Of the 225 houses on an estate, 75 have no garages.
What fraction of the houses (i) have no garage (ii) do have a garage?

e During one week a television network spent 8h on news programmes, 16h on films, 12h on documentaries, 14h on general entertainment and 6h on other programmes.
What fraction of the week's viewing was (i) films (ii) news?

f On an express train from Crewe to Carlisle there are 216 men, 144 women and 120 children.
What fraction of the total number is made up of
(i) women (ii) children (iii) men?

g A newspaper prints 40 columns of news, 14 columns of photographs and 18 columns of advertisements.
What fraction of the newspaper is devoted to (i) news (ii) advertisements?

h Mr Appleby's earnings last week were £81 basic pay, £63 overtime and £45 back pay
What fraction of his wage was (i) basic pay (ii) overtime?

i The villages of Upton, Netherton and Middleton have populations of 350, 126 and 784 respectively.
What fraction of their total population lives in (i) Upton (ii) Netherton?

j Mr Wormlow earned £9720 last year but he paid £3240 in tax and £810 on his mortgage.
What fraction of his earnings went on
(i) tax (ii) his mortgage (iii) other things?

Multiplying and dividing fractions

Introduction

Examples

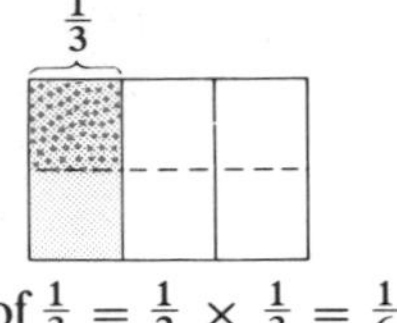

$\frac{1}{2}$ of $\frac{1}{3} = \frac{1}{2} \times \frac{1}{3} = \frac{1}{6}$

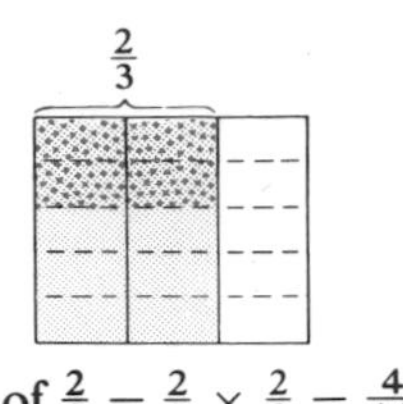

$\frac{2}{5}$ of $\frac{2}{3} = \frac{2}{5} \times \frac{2}{3} = \frac{4}{15}$

The above diagrams illustrate the multiplications shown.

Use the following diagrams to help you multiply the fractions.

1 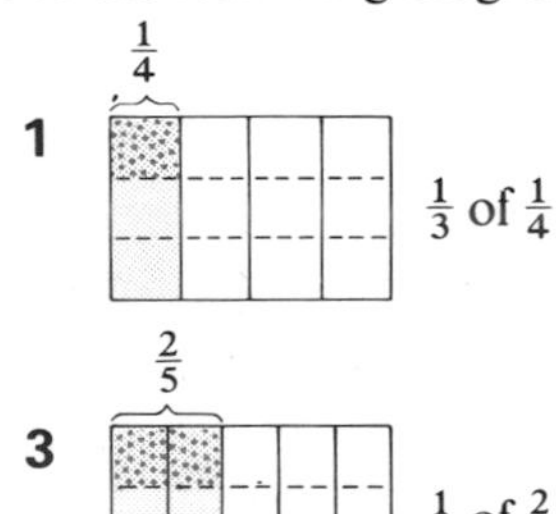 $\frac{1}{3}$ of $\frac{1}{4}$

2 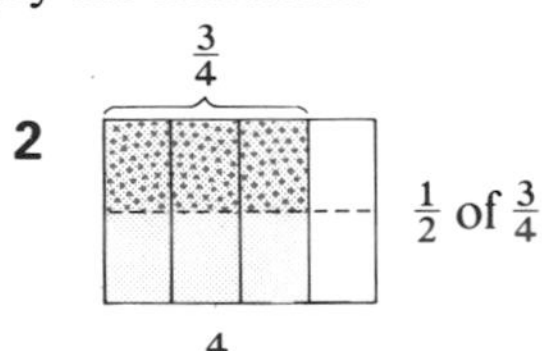 $\frac{1}{2}$ of $\frac{3}{4}$

3 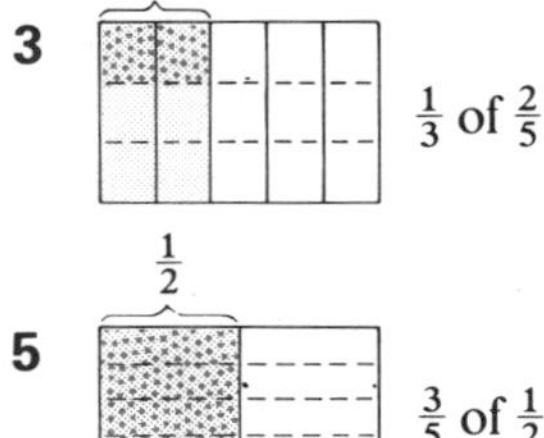 $\frac{1}{3}$ of $\frac{2}{5}$

4 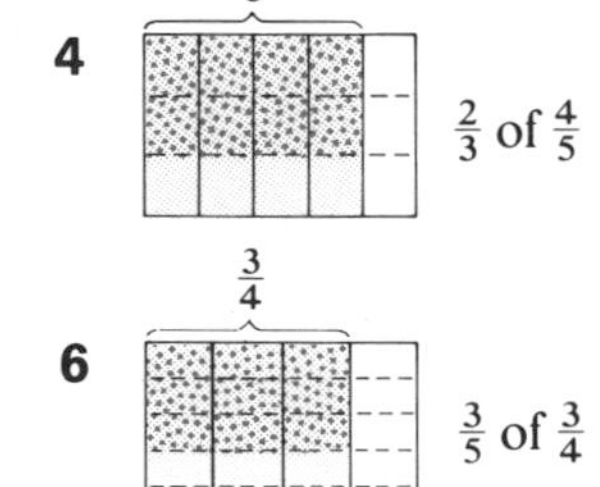 $\frac{2}{3}$ of $\frac{4}{5}$

5 $\frac{3}{5}$ of $\frac{1}{2}$

6 $\frac{3}{5}$ of $\frac{3}{4}$

Part 1 Without cancelling

Work the following.

1 $\frac{3}{5} \times \frac{1}{4}$	**2** $\frac{1}{2} \times \frac{3}{8}$	**3** $\frac{1}{3} \times \frac{2}{3}$	**4** $\frac{2}{3} \times \frac{2}{5}$
5 $\frac{2}{5} \times \frac{1}{3}$	**6** $\frac{2}{7} \times \frac{1}{3}$	**7** $\frac{3}{5} \times \frac{1}{2}$	**8** $\frac{4}{7} \times \frac{1}{3}$
9 $\frac{2}{7} \times \frac{1}{5}$	**10** $\frac{2}{3} \times \frac{2}{7}$	**11** $\frac{4}{5} \times \frac{2}{3}$	**12** $\frac{5}{6} \times \frac{1}{3}$
13 $\frac{3}{4} \times \frac{5}{7}$	**14** $\frac{2}{3} \times \frac{5}{7}$	**15** $\frac{4}{7} \times \frac{2}{3}$	**16** $\frac{5}{6} \times \frac{5}{7}$

Change the mixed numbers into *top-heavy* fractions before multiplying.

17 $1\frac{1}{2} \times \frac{1}{4}$	**18** $1\frac{2}{3} \times \frac{1}{2}$	**19** $1\frac{1}{3} \times \frac{1}{5}$	**20** $\frac{2}{3} \times 1\frac{1}{3}$
21 $\frac{1}{4} \times 1\frac{2}{3}$	**22** $\frac{2}{5} \times 2\frac{1}{3}$	**23** $2\frac{1}{2} \times \frac{1}{4}$	**24** $1\frac{1}{2} \times 2\frac{1}{2}$
25 $1\frac{2}{5} \times 1\frac{1}{3}$	**26** $2\frac{1}{2} \times \frac{3}{4}$	**27** $\frac{1}{7} \times 2\frac{1}{2}$	**28** $1\frac{2}{3} \times 3\frac{1}{2}$

Use the same method when multiplying three or more fractions.

29 $\frac{1}{2} \times \frac{3}{4} \times 1\frac{1}{2}$	**30** $2\frac{1}{2} \times \frac{1}{4} \times 1\frac{1}{2}$	**31** $3\frac{1}{3} \times \frac{1}{3} \times 1\frac{2}{3}$
32 $4\frac{1}{2} \times 2\frac{1}{4} \times \frac{3}{5}$	**33** $3\frac{1}{3} \times \frac{1}{5} \times 1\frac{1}{2}$	**34** $1\frac{2}{11} \times \frac{3}{4} \times \frac{1}{2}$
35 $6\frac{1}{3} \times \frac{1}{3} \times 1\frac{1}{4}$	**36** $\frac{1}{3} \times 2\frac{1}{2} \times 3\frac{1}{4}$	**37** $3\frac{2}{3} \times \frac{1}{2} \times \frac{1}{4}$
38 $5\frac{1}{3} \times 6\frac{1}{9} \times \frac{6}{11}$	**39** $1\frac{1}{2} \times 1\frac{2}{5} \times 2\frac{1}{4}$	**40** $2\frac{1}{5} \times 2\frac{1}{3} \times 3\frac{1}{2}$
41 $\frac{1}{2} \times \frac{3}{4} \times \frac{5}{8}$	**42** $\frac{1}{2} \times 1\frac{1}{2} \times 2\frac{1}{2} \times 3\frac{1}{2} \times 4\frac{1}{2}$	

Multiplying and dividing fractions

43 It takes $2\frac{1}{2}$ hours to paint a fence. One quarter of this time is spent sanding the wood down. What fraction of an hour is spent in sanding the wood?

44 $\frac{3}{5}$ of a road is to be resurfaced. If the road is $1\frac{1}{2}$ km long, what length in kilometres is to be resurfaced?

45 $\frac{2}{3}$ of a man's garden is planted with vegetables. If the area of the garden is $1\frac{1}{3}$ hectares, what area has vegetables in it?

46 I have $1\frac{1}{3}$ litres of oil and I use $\frac{2}{5}$ of it. How many litres have I used?

47 If you spent $1\frac{1}{2}$ hours on homework last night and $\frac{1}{4}$ of that time was on history, what fraction of an hour did you spend on history?

48 A piece of cloth is $3\frac{1}{2}$ metres long. Mrs Barnes uses $\frac{3}{8}$ of it. What length does she use?

49 I have $1\frac{1}{3}$ kg of dog biscuits, and I give my dog $\frac{2}{3}$ of them. What mass of biscuits have I fed it?

50 A man has $3\frac{1}{2}$ hectares on $\frac{3}{5}$ of which he grows apple trees. Yesterday he pruned half of his trees. What area did he prune?

51 Malcolm has $2\frac{1}{3}$ litres of lemonade. I have $2\frac{1}{2}$ times more than he has, and I drink half of mine. How many litres do I drink?

52 The distance to school is $2\frac{1}{2}$ km. The school bus broke down with $\frac{5}{8}$ of the distance left to walk, but my teacher gave me a lift for $\frac{3}{4}$ of this distance. For how many kilometres did my teacher give me a lift?

Part 2 With cancelling

Multiply the fractions, remembering to cancel where possible.

1 $\frac{1}{2} \times \frac{4}{5}$	**2** $\frac{1}{3} \times \frac{6}{7}$	**3** $\frac{2}{5} \times \frac{10}{3}$	**4** $\frac{2}{3} \times \frac{1}{4}$
5 $\frac{6}{5} \times \frac{1}{3}$	**6** $\frac{5}{2} \times \frac{3}{10}$	**7** $\frac{3}{7} \times \frac{2}{6}$	**8** $\frac{4}{5} \times \frac{3}{8}$
9 $\frac{6}{7} \times \frac{5}{12}$	**10** $\frac{3}{5} \times \frac{4}{8}$	**11** $\frac{4}{5} \times \frac{10}{7}$	**12** $\frac{9}{10} \times \frac{11}{18}$
13 $\frac{3}{7} \times \frac{21}{10}$	**14** $\frac{5}{12} \times \frac{6}{7}$	**15** $\frac{2}{3} \times \frac{6}{5} \times \frac{1}{7}$	**16** $\frac{4}{5} \times \frac{3}{7} \times \frac{1}{8}$

Here you can cancel twice.

17 $\frac{3}{7} \times \frac{14}{9}$	**18** $\frac{4}{5} \times \frac{10}{12}$	**19** $\frac{4}{5} \times \frac{15}{24}$	**20** $\frac{6}{18} \times \frac{3}{9}$
21 $\frac{9}{27} \times \frac{8}{4}$	**22** $\frac{10}{30} \times \frac{14}{7}$	**23** $\frac{15}{21} \times \frac{7}{5}$	**24** $\frac{26}{30} \times \frac{5}{13}$

Mixed numbers must be changed into *top-heavy* fractions before cancelling.

25 $\frac{1}{9} \times 1\frac{1}{2}$	**26** $\frac{1}{5} \times 2\frac{1}{2}$	**27** $\frac{2}{7} \times 2\frac{1}{3}$	**28** $1\frac{1}{6} \times \frac{3}{7}$
29 $2\frac{1}{7} \times 1\frac{2}{5}$	**30** $6\frac{9}{10} \times 1\frac{1}{23}$	**31** $\frac{1}{6} \times \frac{18}{64} \times \frac{16}{20}$	**32** $\frac{4}{18} \times \frac{35}{20} \times \frac{9}{14}$

33 $\frac{3}{8} \times 1\frac{7}{9} \times 1\frac{1}{2}$	**34** $\frac{18}{20} \times \frac{140}{36}$	**35** $\frac{16}{36} \times 1\frac{24}{28} \times \frac{3}{8}$
36 $2\frac{1}{7} \times 1\frac{1}{15} \times 22\frac{1}{2} \times \frac{1}{18}$	**37** $6\frac{3}{10} \times 2\frac{1}{7} \times \frac{5}{9}$	**38** $9\frac{2}{3} \times 1\frac{1}{29} \times \frac{6}{15} \times 7\frac{1}{5}$

39 Work the following.

a $\frac{2}{3} \times 12$	**b** $\frac{5}{6} \times 24$	**c** $28 \times \frac{2}{7}$	**d** $20 \times \frac{5}{8}$
e $\frac{7}{10} \times 25$	**f** $\frac{4}{9} \times 15$	**g** $36 \times \frac{3}{8}$	**h** $\frac{1}{2} \times \frac{3}{5} \times 40$
i $\frac{3}{5} \times \frac{3}{4} \times 50$	**j** $\frac{2}{3} \times \frac{7}{8} \times 60$	**k** $\frac{3}{8} \times \frac{4}{5} \times 25$	**l** $\frac{4}{9} \times \frac{3}{8} \times 45$

Multiplying and dividing fractions

40 A rubbish dump covering $7\frac{1}{5}$ hectares is already $\frac{3}{4}$ full. On how many hectares has rubbish been dumped?

41 The area of a small island is $27\frac{1}{2}$ hectares. $\frac{4}{5}$ of the area is rocky. How many hectares is this?

42 A machine component has a mass of $6\frac{1}{4}$ kg. If $\frac{8}{15}$ of it is made from plastic, what mass of plastic has been used?

43 A salesman spends 44 hours in one week travelling by car and train. If $\frac{5}{8}$ of this time is by car, how many hours is this?

44 In my garden I use $\frac{5}{9}$ of a bag of peat with a mass of 30 kg. What mass of peat have I used?

45 There are 216 children in a primary school. $\frac{7}{36}$ of them are in their first year. How many children are
a in their first year **b** not in their first year?

46 A farmer has 1088 fruit trees in an orchard. When $\frac{7}{16}$ of them have been stripped of fruit, how many trees have
a no fruit on them **b** some fruit still on them?

47 Another farmer spends $\frac{5}{7}$ of one month's expenditure on animal foodstuff. Of this foodstuff, $\frac{3}{4}$ is for his cows. If he spends £336 that month, how much does he spend on his cows?

48 A man has a 72 kg bag of cement. His first job uses $\frac{3}{4}$ of it. Of what is left, his second job uses $\frac{1}{3}$. Then, of the remainder, the third job uses $\frac{1}{2}$. What mass is now left? What fraction is it of the original bag?

49 $\frac{3}{4}$ of the total number of girls in a school play hockey. Of those who play hockey, $\frac{1}{2}$ also play tennis. Of those who play hockey and tennis, $\frac{3}{5}$ also play netball. If there are 400 girls in the school, how many play all three games?

50 In a village of 420 people, $\frac{2}{5}$ are children under 16 years old. Of these children, $\frac{4}{7}$ are boys; and of these boys $\frac{1}{3}$ are under 7 years old. If each boy under 7 years old is given a $\frac{3}{4}$ kg box of chocolates, what mass of chocolates is required?

Part 3 Dividing by a whole number

For example, dividing a number by 2 has the same effect as halving it.
So $\frac{1}{3} \div 2 = \frac{1}{3} \times \frac{1}{2} = \frac{1}{6}$

Work the following.

1 $\frac{1}{4} \div 2$ **2** $\frac{1}{5} \div 2$ **3** $\frac{1}{4} \div 3$ **4** $\frac{1}{2} \div 3$
5 $\frac{2}{3} \div 3$ **6** $\frac{3}{4} \div 2$ **7** $\frac{3}{5} \div 2$ **8** $\frac{4}{5} \div 5$
9 $\frac{3}{7} \div 2$ **10** $\frac{5}{8} \div 4$ **11** $\frac{7}{8} \div 3$ **12** $\frac{4}{9} \div 5$
13 $\frac{7}{10} \div 3$ **14** $\frac{1}{2} \div 8$ **15** $\frac{3}{4} \div 4$ **16** $\frac{5}{12} \div 2$

Change mixed numbers into *top-heavy* fractions before dividing.

17 $1\frac{1}{2} \div 2$ **18** $2\frac{1}{2} \div 3$ **19** $4\frac{1}{2} \div 5$ **20** $1\frac{1}{3} \div 3$
21 $2\frac{2}{3} \div 5$ **22** $1\frac{3}{4} \div 2$ **23** $3\frac{1}{4} \div 4$ **24** $2\frac{1}{5} \div 3$
25 $4\frac{2}{3} \div 5$ **26** $1\frac{3}{8} \div 4$ **27** $5\frac{1}{5} \div 7$ **28** $4\frac{7}{8} \div 2$

Multiplying and dividing fractions

Work the following. Remember to cancel where possible.

29 $\frac{4}{5} \div 2$ **30** $\frac{6}{7} \div 3$ **31** $\frac{8}{9} \div 4$ **32** $\frac{4}{7} \div 4$

33 $\frac{9}{11} \div 3$ **34** $\frac{2}{3} \div 6$ **35** $\frac{4}{5} \div 8$ **36** $\frac{3}{4} \div 9$

37 $\frac{5}{6} \div 20$ **38** $\frac{6}{7} \div 9$ **39** $\frac{4}{9} \div 6$ **40** $\frac{9}{10} \div 12$

41 $\frac{8}{11} \div 10$ **42** $\frac{4}{5} \div 6$ **43** $\frac{8}{15} \div 4$ **44** $\frac{6}{7} \div 3$

Change the mixed numbers into *top-heavy* fractions before dividing. Cancel where possible.

45 $4\frac{1}{2} \div 6$ **46** $7\frac{1}{2} \div 20$ **47** $2\frac{2}{3} \div 4$ **48** $3\frac{1}{3} \div 5$

49 $1\frac{1}{3} \div 8$ **50** $1\frac{1}{4} \div 10$ **51** $2\frac{2}{3} \div 12$ **52** $2\frac{1}{2} \div 10$

53 $1\frac{1}{5} \div 9$ **54** $4\frac{2}{3} \div 6$ **55** $4\frac{1}{2} \div 3$ **56** $7\frac{1}{2} \div 5$

57 $12\frac{1}{2} \div 5$ **58** $2\frac{2}{3} \div 2$ **59** $3\frac{1}{3} \div 2$ **60** $5\frac{1}{4} \div 3$

61 $3\frac{1}{2} \div 2$ **62** $4\frac{2}{3} \div 4$ **63** $6\frac{3}{4} \div 6$ **64** $22\frac{1}{2} \div 12$

65 $19\frac{1}{2} \div 15$ **66** $3\frac{1}{3} \div 12$ **67** $12\frac{3}{4} \div 6$ **68** $7\frac{1}{2} \div 25$

Part 4 Dividing by a fraction

Work the following.

1 $\frac{2}{5} \div \frac{1}{2}$ **2** $\frac{3}{7} \div \frac{1}{2}$ **3** $\frac{2}{9} \div \frac{1}{2}$ **4** $\frac{3}{11} \div \frac{1}{3}$

5 $\frac{1}{2} \div \frac{2}{3}$ **6** $\frac{1}{4} \div \frac{2}{5}$ **7** $\frac{2}{5} \div \frac{3}{4}$ **8** $\frac{3}{8} \div \frac{2}{3}$

Change mixed numbers into *top-heavy* fractions before dividing.

9 $1\frac{1}{2} \div 1\frac{3}{5}$ **10** $2\frac{1}{2} \div 2\frac{2}{3}$ **11** $1\frac{1}{3} \div 1\frac{1}{2}$ **12** $4\frac{1}{2} \div 5\frac{1}{3}$

13 $3\frac{1}{3} \div 3\frac{1}{2}$ **14** $2\frac{1}{4} \div 2\frac{2}{3}$ **15** $2\frac{3}{5} \div 3\frac{1}{2}$ **16** $5\frac{1}{2} \div 6\frac{1}{3}$

17 $1\frac{1}{2} \div \frac{1}{3}$ **18** $2\frac{1}{2} \div \frac{1}{3}$ **19** $4\frac{1}{2} \div 1\frac{1}{3}$ **20** $3\frac{1}{2} \div 2\frac{2}{3}$

21 $5\frac{1}{2} \div 1\frac{1}{3}$ **22** $2\frac{3}{4} \div \frac{4}{5}$ **23** $4\frac{3}{5} \div \frac{2}{3}$ **24** $2\frac{1}{5} \div 1\frac{1}{3}$

Cancel where possible.

25 $\frac{3}{10} \div \frac{1}{2}$ **26** $\frac{1}{6} \div \frac{2}{3}$ **27** $\frac{4}{9} \div \frac{2}{3}$ **28** $\frac{7}{10} \div \frac{2}{5}$

29 $\frac{3}{8} \div \frac{1}{4}$ **30** $\frac{5}{12} \div \frac{1}{6}$ **31** $\frac{6}{7} \div \frac{3}{5}$ **32** $\frac{5}{9} \div \frac{5}{6}$

Change mixed numbers into *top-heavy* fractions and cancel where possible.

33 $3\frac{1}{3} \div 7\frac{1}{2}$ **34** $4\frac{1}{2} \div 7\frac{1}{2}$ **35** $1\frac{4}{5} \div 2\frac{7}{10}$ **36** $1\frac{1}{3} \div 2\frac{2}{9}$

37 $4\frac{4}{5} \div 5\frac{1}{3}$ **38** $3\frac{1}{9} \div 4\frac{2}{3}$ **39** $6\frac{2}{5} \div 1\frac{1}{15}$ **40** $4\frac{2}{3} \div 1\frac{1}{6}$

41 $\dfrac{2\frac{2}{3}}{2\frac{2}{5}}$ **42** $\dfrac{10\frac{1}{2}}{2\frac{1}{4}}$ **43** $\dfrac{4\frac{1}{2}}{3\frac{3}{4}}$ **44** $\dfrac{4\frac{2}{3}}{3\frac{1}{9}}$ **45** $\dfrac{2\frac{4}{7}}{2\frac{2}{5}}$

46 $\dfrac{2\frac{1}{4}}{1\frac{7}{8}}$ **47** $\dfrac{12\frac{1}{2}}{4\frac{3}{8}}$ **48** $\dfrac{2\frac{2}{3}}{2\frac{2}{9}}$ **49** $\dfrac{11\frac{2}{3}}{2\frac{7}{9}}$ **50** $\dfrac{13\frac{1}{2}}{5\frac{1}{4}}$

Multiplying and dividing fractions

51 Mrs Hickman bought $6\frac{2}{3}$ metres of dress material and used $1\frac{2}{3}$ metres of it. What fraction of the material has she used?

52 A plank of wood $6\frac{2}{5}$ metres long has a length of $2\frac{2}{5}$ metres cut off. What fraction of the plank is the piece which is cut off?

53 A sheet of metal of area $20\frac{1}{4}$ cm^2 has a hole of area $4\frac{1}{2}$ cm^2 punched in it. What fraction of the sheet is the hole?

54 A photograph has an area of $17\frac{1}{2}$ cm^2 of which $5\frac{1}{4}$ cm^2 is border. What fraction of the photograph is the border?

55 Mr Gibson had a mass of $16\frac{1}{2}$ stone. During a diet he lost enough mass to bring him down to $14\frac{2}{3}$ stone. What fraction of his old mass is his new mass?

56 A man set out on a journey of $67\frac{1}{2}$ miles and had a puncture after only $13\frac{1}{2}$ miles. What fraction of his journey had he covered before his puncture?

57 A farmer has a field of $21\frac{2}{3}$ hectares. If he ploughs $6\frac{1}{2}$ hectares of the field, what fraction of the field is ploughed?

58 A man sleeps for $7\frac{1}{2}$ hours every 24 hours. For what fraction of the day is the man asleep?

59 A $17\frac{1}{2}$-litre tank of oil has 10 litres poured out. What fraction of the oil has been poured out?

60 $13\frac{1}{2}$ kg of coffee is put into $\frac{3}{4}$ kg packets. How many packets will be needed?

61 A tin holds $10\frac{2}{3}$ litres of washing-up liquid. How many times will it fill a container holding $\frac{2}{3}$ litre?

62 A metal sheet is $12\frac{3}{5}$ metres long. How many short strips $\frac{3}{10}$ metres long can be cut from this sheet?

63 A road is $28\frac{4}{5}$ km long and it is being resurfaced at a rate of $3\frac{1}{5}$ km each day. How many days will resurfacing take?

64 A man owns $46\frac{2}{3}$ hectares of land which he wants to plough. If he ploughs $2\frac{2}{9}$ hectares each day, how long will the ploughing take?

65 I have $16\frac{1}{5}$ kg of meal to feed animals. If I use $2\frac{7}{10}$ kg each day, how long will the meal last?

Halves, quarters, eighths etc.

Equivalent fractions

Copy and complete the following.

$\frac{1}{2} = \frac{\ }{4} = \frac{\ }{8} = \frac{\ }{16} = \frac{\ }{32}$ $\frac{1}{4} = \frac{\ }{8} = \frac{\ }{16} = \frac{\ }{32}$ $\frac{3}{4} = \frac{\ }{8} = \frac{\ }{16} = \frac{\ }{32}$ $\frac{1}{8} = \frac{\ }{16} = \frac{\ }{32}$

$\frac{3}{8} = \frac{\ }{16} = \frac{\ }{32}$ $\frac{5}{8} = \frac{\ }{16} = \frac{\ }{32}$ $\frac{7}{8} = \frac{\ }{16} = \frac{\ }{32}$ $\frac{9}{16} = \frac{\ }{32}$

$\frac{3}{16} = \frac{\ }{32}$ $\frac{7}{16} = \frac{\ }{32}$ $\frac{5}{16} = \frac{\ }{32}$ $\frac{15}{16} = \frac{\ }{32}$ $\frac{11}{16} = \frac{\ }{32}$ $\frac{13}{16} = \frac{\ }{32}$

Addition

Part 1

1 $\frac{1}{2} + \frac{1}{4}$ 2 $\frac{1}{2} + \frac{1}{8}$ 3 $\frac{1}{2} + \frac{3}{8}$ 4 $\frac{1}{4} + \frac{1}{8}$ 5 $\frac{1}{4} + \frac{3}{8}$

6 $\frac{3}{4} + \frac{1}{8}$ 7 $\frac{1}{2} + \frac{1}{16}$ 8 $\frac{1}{2} + \frac{3}{16}$ 9 $\frac{1}{4} + \frac{1}{16}$ 10 $\frac{1}{4} + \frac{3}{16}$

11 $\frac{3}{4} + \frac{1}{16}$ 12 $\frac{3}{4} + \frac{3}{16}$ 13 $\frac{1}{8} + \frac{3}{16}$ 14 $\frac{3}{8} + \frac{1}{16}$ 15 $\frac{3}{8} + \frac{1}{4}$

16 $\frac{1}{8} + \frac{5}{32}$ 17 $\frac{5}{8} + \frac{1}{32}$ 18 $\frac{5}{16} + \frac{1}{32}$ 19 $\frac{7}{16} + \frac{3}{32}$ 20 $\frac{9}{16} + \frac{5}{32}$

21 $\frac{1}{2} + \frac{3}{32}$ 22 $\frac{1}{4} + \frac{5}{32}$ 23 $\frac{1}{4} + \frac{11}{32}$ 24 $\frac{3}{4} + \frac{1}{32}$ 25 $\frac{5}{16} + \frac{5}{32}$

26 $\frac{3}{4} + \frac{7}{32}$ 27 $\frac{1}{2} + \frac{15}{32}$ 28 $\frac{7}{16} + \frac{1}{4}$ 29 $\frac{3}{8} + \frac{3}{32}$ 30 $\frac{1}{8} + \frac{5}{16}$

Part 2

1 $\frac{1}{2} + \frac{3}{4}$ 2 $\frac{1}{2} + \frac{5}{8}$ 3 $\frac{1}{2} + \frac{7}{8}$ 4 $\frac{1}{4} + \frac{7}{8}$ 5 $\frac{3}{4} + \frac{5}{8}$

6 $\frac{3}{4} + \frac{3}{8}$ 7 $\frac{3}{4} + \frac{7}{8}$ 8 $\frac{1}{2} + \frac{15}{16}$ 9 $\frac{1}{8} + \frac{15}{16}$ 10 $\frac{1}{4} + \frac{15}{16}$

11 $\frac{3}{4} + \frac{5}{16}$ 12 $\frac{1}{4} + \frac{13}{16}$ 13 $\frac{3}{8} + \frac{11}{16}$ 14 $\frac{7}{8} + \frac{3}{16}$ 15 $\frac{1}{2} + \frac{9}{16}$

16 $\frac{5}{8} + \frac{13}{16}$ 17 $\frac{9}{16} + \frac{5}{8}$ 18 $\frac{1}{2} + \frac{3}{4} + \frac{7}{8}$ 19 $\frac{3}{8} + \frac{3}{16} + \frac{5}{32}$

20 $\frac{1}{2} + \frac{1}{4} + \frac{1}{8} + \frac{1}{16} + \frac{1}{32}$ 21 $\frac{1}{2} + \frac{3}{4} + \frac{5}{8} + \frac{7}{16} + \frac{9}{32}$

22 $\frac{1}{2} + \frac{2}{4} + \frac{3}{8} + \frac{4}{16} + \frac{5}{32}$

Part 3

1 $\frac{1}{2} + \frac{15}{16}$ 2 $\frac{15}{16} + \frac{7}{8}$ 3 $\frac{3}{4} + \frac{31}{32}$ 4 $\frac{9}{16} + \frac{1}{2}$ 5 $\frac{3}{4} + \frac{9}{16}$

6 $\frac{7}{8} + \frac{11}{32}$ 7 $\frac{9}{16} + \frac{15}{32}$ 8 $\frac{7}{8} + \frac{11}{16}$ 9 $\frac{17}{32} + \frac{5}{8}$ 10 $\frac{9}{16} + \frac{19}{32}$

11 $\frac{15}{16} + \frac{21}{32}$ 12 $\frac{3}{4} + \frac{15}{16}$ 13 $\frac{1}{2} + \frac{3}{4} + \frac{7}{8}$ 14 $\frac{5}{8} + \frac{5}{16} + \frac{5}{32}$

15 $\frac{29}{32} + \frac{9}{16} + \frac{1}{4}$ 16 $\frac{7}{8} + \frac{7}{16} + \frac{7}{32}$ 17 $\frac{13}{16} + \frac{9}{32} + \frac{1}{2}$

18 $\frac{1}{2} + \frac{3}{4} + \frac{7}{8} + \frac{15}{16}$ 19 $\frac{1}{2} + \frac{3}{4} + \frac{7}{8} + \frac{15}{16} + \frac{31}{32}$ 20 $\frac{1}{2} + \frac{19}{32} + \frac{15}{16} + \frac{5}{8}$

Subtraction

Part 4

1 $1 - \frac{1}{4}$ 2 $1 - \frac{1}{8}$ 3 $1 - \frac{3}{4}$ 4 $1 - \frac{3}{8}$ 5 $1 - \frac{1}{16}$

6 $1 - \frac{15}{16}$ 7 $2 - \frac{1}{4}$ 8 $3 - \frac{1}{8}$ 9 $3 - \frac{5}{8}$ 10 $4 - \frac{7}{8}$

11 $2 - \frac{1}{8}$ 12 $2 - \frac{3}{4}$ 13 $6 - \frac{1}{2}$ 14 $6 - \frac{15}{16}$ 15 $6 - \frac{7}{8}$

16 $1 - \frac{5}{8}$ 17 $1 - \frac{11}{16}$ 18 $1 - \frac{31}{32}$ 19 $1 - \frac{1}{32}$ 20 $2 - \frac{19}{32}$

21 $3 - \frac{17}{32}$ 22 $7 - \frac{7}{8}$ 23 $6 - \frac{11}{16}$ 24 $3 - \frac{13}{16}$

Halves, quarters, eighths etc.

Part 5

1 $\frac{3}{4} - \frac{1}{2}$ 2 $\frac{1}{2} - \frac{1}{4}$ 3 $\frac{1}{4} - \frac{1}{8}$ 4 $\frac{1}{2} - \frac{1}{8}$ 5 $\frac{1}{2} - \frac{3}{8}$

6 $\frac{3}{4} - \frac{1}{8}$ 7 $\frac{3}{4} - \frac{3}{8}$ 8 $\frac{3}{4} - \frac{5}{8}$ 9 $\frac{7}{8} - \frac{1}{4}$ 10 $\frac{7}{8} - \frac{1}{2}$

11 $\frac{7}{8} - \frac{3}{4}$ 12 $\frac{5}{8} - \frac{1}{4}$ 13 $\frac{5}{8} - \frac{1}{2}$ 14 $\frac{15}{16} - \frac{1}{2}$ 15 $\frac{9}{16} - \frac{1}{2}$

16 $\frac{5}{16} - \frac{1}{4}$ 17 $\frac{13}{16} - \frac{3}{4}$ 18 $\frac{7}{8} - \frac{1}{16}$ 19 $\frac{7}{8} - \frac{13}{16}$ 20 $\frac{3}{4} - \frac{3}{16}$

21 $\frac{13}{16} - \frac{1}{2}$ 22 $\frac{31}{32} - \frac{1}{2}$ 23 $\frac{17}{32} - \frac{1}{2}$ 24 $\frac{15}{32} - \frac{1}{4}$ 25 $\frac{15}{16} - \frac{3}{4}$

26 $\frac{7}{8} - \frac{27}{32}$ 27 $\frac{5}{8} - \frac{1}{16}$ 28 $\frac{25}{32} - \frac{3}{4}$ 29 $\frac{3}{4} - \frac{7}{32}$ 30 $\frac{1}{4} - \frac{3}{16}$

Part 6

1 $3\frac{1}{2} - 1\frac{1}{4}$ 2 $5\frac{3}{4} - 2\frac{1}{2}$ 3 $6\frac{7}{8} - 1\frac{1}{2}$ 4 $3\frac{5}{8} - 2\frac{1}{2}$

5 $4\frac{3}{4} - 1\frac{3}{8}$ 6 $2\frac{1}{2} - 1\frac{3}{8}$ 7 $5\frac{1}{2} - 1\frac{7}{16}$ 8 $3\frac{3}{8} - 1\frac{5}{16}$

9 $4\frac{1}{4} - 3\frac{3}{16}$ 10 $7\frac{3}{4} - 6\frac{11}{16}$ 11 $5\frac{7}{8} - 4\frac{11}{16}$ 12 $3\frac{3}{4} - 1\frac{9}{16}$

13 $2\frac{7}{8} - 1\frac{1}{2}$ 14 $8\frac{5}{8} - 4\frac{3}{16}$ 15 $7\frac{3}{8} - 3\frac{3}{16}$ 16 $9\frac{3}{4} - 8\frac{23}{32}$

17 $7\frac{1}{2} - 6\frac{15}{32}$ 18 $4\frac{1}{4} - 4\frac{5}{32}$

A mixture

Part 7

	a	b	c
1	$\frac{3}{4} \times \frac{1}{2}$	$\frac{3}{4} + \frac{1}{2}$	$\frac{3}{4} - \frac{1}{2}$
2	$\frac{1}{2} \times \frac{3}{8}$	$\frac{1}{2} + \frac{3}{8}$	$\frac{1}{2} - \frac{3}{8}$
3	$\frac{1}{2} \times \frac{1}{8}$	$\frac{1}{2} + \frac{1}{8}$	$\frac{1}{2} - \frac{1}{8}$
4	$\frac{1}{4} \times \frac{1}{8}$	$\frac{1}{4} + \frac{1}{8}$	$\frac{1}{4} - \frac{1}{8}$
5	$\frac{3}{8} \times \frac{1}{4}$	$\frac{3}{8} + \frac{1}{4}$	$\frac{3}{8} - \frac{1}{4}$
6	$\frac{3}{4} \times \frac{1}{8}$	$\frac{3}{4} + \frac{1}{8}$	$\frac{3}{4} - \frac{1}{8}$
7	$\frac{7}{8} \times \frac{1}{2}$	$\frac{7}{8} + \frac{1}{2}$	$\frac{7}{8} - \frac{1}{2}$
8	$\frac{1}{2} \times \frac{5}{16}$	$\frac{1}{2} + \frac{5}{16}$	$\frac{1}{2} - \frac{5}{16}$
9	$\frac{3}{4} \times \frac{3}{16}$	$\frac{3}{4} + \frac{3}{16}$	$\frac{3}{4} - \frac{3}{16}$
10	$\frac{9}{16} \times \frac{1}{2}$	$\frac{9}{16} + \frac{1}{2}$	$\frac{9}{16} - \frac{1}{2}$
11	$\frac{3}{4} \times \frac{5}{8}$	$\frac{3}{4} + \frac{5}{8}$	$\frac{3}{4} - \frac{5}{8}$
12	$\frac{7}{8} \times \frac{1}{2}$	$\frac{7}{8} + \frac{1}{2}$	$\frac{7}{8} - \frac{1}{2}$
13	$\frac{7}{8} \times \frac{3}{4}$	$\frac{7}{8} + \frac{3}{4}$	$\frac{7}{8} - \frac{3}{4}$
14	$\frac{3}{4} \times \frac{9}{16}$	$\frac{3}{4} + \frac{9}{16}$	$\frac{3}{4} - \frac{9}{16}$
15	$\frac{11}{32} \times \frac{1}{4}$	$\frac{11}{32} + \frac{1}{4}$	$\frac{11}{32} - \frac{1}{4}$

Part 8

	a	b	c
1	$1\frac{1}{2} \times \frac{1}{2}$	$1\frac{1}{2} + \frac{1}{2}$	$1\frac{1}{2} - \frac{1}{2}$
2	$2\frac{1}{2} \times \frac{1}{4}$	$2\frac{1}{2} + \frac{1}{4}$	$2\frac{1}{2} - \frac{1}{4}$
3	$2\frac{3}{4} \times \frac{1}{4}$	$2\frac{3}{4} + \frac{1}{4}$	$2\frac{3}{4} - \frac{1}{4}$
4	$1\frac{1}{4} \times \frac{1}{2}$	$1\frac{1}{4} + \frac{1}{2}$	$1\frac{1}{4} - \frac{1}{2}$
5	$3\frac{1}{2} \times \frac{1}{8}$	$3\frac{1}{2} + \frac{1}{8}$	$3\frac{1}{2} - \frac{1}{8}$
6	$1\frac{1}{2} \times 1\frac{1}{4}$	$1\frac{1}{2} + 1\frac{1}{4}$	$1\frac{1}{2} - 1\frac{1}{4}$
7	$2\frac{1}{2} \times 1\frac{1}{4}$	$2\frac{1}{2} + 1\frac{1}{4}$	$2\frac{1}{2} - 1\frac{1}{4}$
8	$1\frac{7}{8} \times \frac{3}{4}$	$1\frac{7}{8} + \frac{3}{4}$	$1\frac{7}{8} - \frac{3}{4}$
9	$5\frac{1}{2} \times \frac{1}{8}$	$5\frac{1}{2} + \frac{1}{8}$	$5\frac{1}{2} - \frac{1}{8}$
10	$2\frac{5}{8} \times 1\frac{1}{2}$	$2\frac{5}{8} + 1\frac{1}{2}$	$2\frac{5}{8} - 1\frac{1}{2}$
11	$(1\frac{1}{2})^2$	$1\frac{1}{2} + 1\frac{1}{2}$	$1\frac{1}{2} - 1\frac{1}{2}$
12	$(2\frac{1}{2})^2$	$2\frac{1}{2} + 2\frac{1}{2}$	$2\frac{1}{2} - 2\frac{1}{2}$
13	$3\frac{1}{2} \times 1\frac{3}{4}$	$3\frac{1}{2} + 1\frac{3}{4}$	$3\frac{1}{2} - 1\frac{3}{4}$
14	$5\frac{7}{8} \times 1\frac{1}{2}$	$5\frac{7}{8} + 1\frac{1}{2}$	$5\frac{7}{8} - 1\frac{1}{2}$
15	$3\frac{3}{4} \times 2\frac{1}{8}$	$3\frac{3}{4} + 2\frac{1}{8}$	$3\frac{3}{4} - 2\frac{1}{8}$
16	$4\frac{1}{8} \times 1\frac{3}{4}$	$4\frac{1}{8} + 1\frac{3}{4}$	$4\frac{1}{8} - 1\frac{3}{4}$

Addition of fractions

Part 1

Work the following.

1 $\frac{3}{4} + \frac{1}{4}$ 2 $\frac{5}{8} + \frac{3}{8}$ 3 $\frac{3}{5} + \frac{2}{5}$ 4 $\frac{7}{10} + \frac{3}{10}$

5 $\frac{5}{8} + \frac{4}{8}$ 6 $\frac{4}{5} + \frac{2}{5}$ 7 $\frac{4}{7} + \frac{5}{7}$ 8 $\frac{7}{9} + \frac{4}{9}$

9 $\frac{3}{4} + \frac{3}{4} + \frac{1}{4}$ 10 $\frac{3}{8} + \frac{7}{8} + \frac{5}{8}$ 11 $\frac{3}{5} + \frac{4}{5} + \frac{4}{5}$ 12 $\frac{6}{7} + \frac{5}{7} + \frac{6}{7}$

13 $3\frac{2}{5} + 4\frac{1}{5}$ 14 $5\frac{3}{8} + 2\frac{2}{8}$ 15 $8\frac{2}{7} + 1\frac{3}{7}$ 16 $4\frac{3}{4} + 2\frac{1}{4}$

17 $6\frac{2}{3} + 2\frac{1}{3}$ 18 $1\frac{7}{8} + 3\frac{1}{8}$ 19 $2\frac{2}{5} + 4\frac{3}{5}$ 20 $5\frac{2}{8} + 2\frac{7}{8}$

21 $3\frac{4}{7} + 6\frac{5}{7}$ 22 $2\frac{5}{9} + 1\frac{5}{9}$ 23 $4\frac{7}{9} + 2\frac{4}{9}$ 24 $8\frac{2}{7} + 3\frac{6}{7}$

25 $4\frac{3}{4} + 2\frac{1}{4} + 5\frac{3}{4}$ 26 $3\frac{3}{8} + 2\frac{7}{8} + 4\frac{5}{8}$ 27 $6\frac{6}{7} + \frac{3}{7} + 2\frac{6}{7}$

28 $\frac{7}{9} + 1\frac{8}{9} + 2\frac{7}{9}$ 29 $8\frac{2}{3} + 4\frac{2}{3} + \frac{1}{3}$ 30 $4\frac{7}{8} + \frac{5}{8} + 2\frac{7}{8}$

Part 2

Copy and complete the following.

1 $\frac{3}{4} + \frac{1}{8} = \frac{}{8} + \frac{1}{8} = \frac{}{8}$ 2 $\frac{2}{3} + \frac{2}{9} = \frac{}{9} + \frac{2}{9} = \frac{}{9}$

3 $\frac{2}{5} + \frac{7}{15} = \frac{}{15} + \frac{7}{15} = \frac{}{15}$ 4 $\frac{3}{8} + \frac{5}{16} = \frac{}{16} + \frac{5}{16} = \frac{}{16}$

5 $\frac{1}{3} + \frac{7}{12} = \frac{}{12} + \frac{7}{12} = \frac{}{12}$ 6 $\frac{1}{5} + \frac{7}{10} = \frac{}{10} + \frac{7}{10} = \frac{}{10}$

7 $\frac{2}{3} + \frac{7}{24} = \frac{}{24} + \frac{7}{24} = \frac{}{24}$ 8 $\frac{1}{15} + \frac{1}{5} = \frac{1}{15} + \frac{}{15} = \frac{}{15}$

9 $\frac{9}{20} + \frac{2}{5} = \frac{9}{20} + \frac{}{20} = \frac{}{20}$ 10 $\frac{4}{9} + \frac{1}{3} = \frac{4}{9} + \frac{}{9} = \frac{}{9}$

Work these additions.

11 $\frac{3}{4} + \frac{1}{16}$ 12 $\frac{5}{8} + \frac{1}{16}$ 13 $\frac{1}{2} + \frac{3}{8}$ 14 $\frac{3}{5} + \frac{1}{10}$

15 $\frac{5}{6} + \frac{1}{12}$ 16 $\frac{1}{5} + \frac{13}{20}$ 17 $\frac{2}{3} + \frac{1}{24}$ 18 $\frac{2}{5} + \frac{7}{30}$

Cancel your answer where possible.

19 $\frac{2}{3} + \frac{1}{12}$ 20 $\frac{2}{5} + \frac{4}{15}$ 21 $\frac{3}{4} + \frac{1}{12}$ 22 $\frac{1}{2} + \frac{1}{8}$

23 $\frac{1}{4} + \frac{5}{12}$ 24 $\frac{1}{3} + \frac{1}{6}$ 25 $\frac{3}{5} + \frac{1}{20}$ 26 $\frac{1}{4} + \frac{7}{24}$

27 $\frac{2}{7} + \frac{3}{14}$ 28 $\frac{3}{8} + \frac{5}{24}$ 29 $\frac{5}{6} + \frac{1}{18}$ 30 $\frac{3}{4} + \frac{1}{16}$

Change *top-heavy* fractions to mixed numbers.

31 $\frac{1}{2} + \frac{3}{4}$ 32 $\frac{3}{4} + \frac{5}{8}$ 33 $\frac{2}{3} + \frac{5}{6}$ 34 $\frac{1}{2} + \frac{5}{8}$

35 $\frac{3}{5} + \frac{7}{10}$ 36 $\frac{8}{9} + \frac{2}{3}$ 37 $\frac{3}{8} + \frac{1}{4}$ 38 $\frac{4}{9} + \frac{1}{3}$

39 $\frac{2}{3} + \frac{7}{12}$ 40 $\frac{3}{4} + \frac{5}{12}$ 41 $\frac{5}{6} + \frac{11}{12}$ 42 $\frac{1}{2} + \frac{1}{12}$

43 $3\frac{1}{2} + 2\frac{7}{8}$ 44 $5\frac{3}{4} + 2\frac{3}{8}$ 45 $1\frac{3}{5} + 2\frac{9}{10}$ 46 $6\frac{2}{3} + 2\frac{7}{9}$

47 $8\frac{3}{4} + 1\frac{7}{12}$ 48 $3\frac{7}{9} + 2\frac{2}{3}$ 49 $5\frac{1}{2} + 1\frac{11}{12}$ 50 $8\frac{4}{5} + 3\frac{7}{15}$

51 $9\frac{7}{8} + \frac{9}{16}$ 52 $3\frac{5}{16} + 2\frac{3}{4}$ 53 $8\frac{3}{7} + \frac{9}{14}$ 54 $\frac{15}{32} + 4\frac{1}{2}$

Addition of fractions

Part 3

Copy and complete the following.

1 $\frac{1}{4} + \frac{2}{3} = \frac{\ }{12} + \frac{\ }{12} = \frac{\ }{12}$

2 $\frac{2}{5} + \frac{1}{6} = \frac{\ }{30} + \frac{\ }{30} = \frac{\ }{30}$

3 $\frac{1}{2} + \frac{4}{9} = \frac{\ }{18} + \frac{\ }{18} = \frac{\ }{18}$

4 $\frac{2}{3} + \frac{1}{5} = \frac{\ }{15} + \frac{\ }{15} = \frac{\ }{15}$

5 $\frac{1}{6} + \frac{3}{8} = \frac{\ }{24} + \frac{\ }{24} = \frac{\ }{24}$

6 $\frac{3}{4} + \frac{1}{6} = \frac{\ }{12} + \frac{\ }{12} = \frac{\ }{12}$

7 $\frac{2}{5} + \frac{3}{7} = \frac{\ }{35} + \frac{\ }{35} = \frac{\ }{35}$

8 $\frac{1}{8} + \frac{2}{3} = \frac{\ }{24} + \frac{\ }{24} = \frac{\ }{24}$

9 $\frac{4}{9} + \frac{2}{5} = \frac{\ }{45} + \frac{\ }{45} = \frac{\ }{45}$

10 $\frac{2}{7} + \frac{3}{5} = \frac{\ }{35} + \frac{\ }{35} = \frac{\ }{35}$

Work these additions.

11 $\frac{1}{2} + \frac{1}{3}$	12 $\frac{1}{3} + \frac{2}{5}$	13 $\frac{3}{4} + \frac{1}{5}$	14 $\frac{2}{3} + \frac{1}{4}$
15 $\frac{2}{5} + \frac{2}{9}$	16 $\frac{2}{7} + \frac{3}{8}$	17 $\frac{1}{2} + \frac{1}{5}$	18 $\frac{3}{4} + \frac{1}{6}$

Change *top-heavy* fractions to mixed numbers. Cancel where possible.

19 $\frac{2}{3} + \frac{1}{2}$	20 $\frac{3}{4} + \frac{2}{5}$	21 $\frac{3}{5} + \frac{1}{2}$	22 $\frac{2}{3} + \frac{3}{4}$
23 $\frac{1}{2} + \frac{4}{5}$	24 $\frac{5}{6} + \frac{1}{5}$	25 $\frac{5}{9} + \frac{3}{4}$	26 $\frac{7}{10} + \frac{2}{3}$
27 $2\frac{3}{4} + 3\frac{2}{3}$	28 $1\frac{5}{6} + 4\frac{1}{2}$	29 $2\frac{3}{5} + 3\frac{1}{4}$	30 $5\frac{5}{6} + 2\frac{2}{3}$
31 $7\frac{3}{8} + 1\frac{3}{4}$	32 $8\frac{3}{10} + 1\frac{2}{5}$	33 $2\frac{8}{9} + 1\frac{1}{4}$	34 $4\frac{2}{3} + 3\frac{3}{5}$
35 $5\frac{1}{2} + 6\frac{2}{7}$	36 $\frac{7}{12} + 3\frac{3}{4}$	37 $\frac{8}{9} + 4\frac{2}{3}$	38 $2\frac{5}{8} + 9\frac{1}{2}$
39 $6\frac{5}{12} + 4\frac{3}{8}$	40 $2\frac{5}{6} + 3\frac{11}{15}$		

41 I post two parcels with masses of $3\frac{3}{4}$ kg and $2\frac{1}{2}$ kg. What total mass do I post?

42 Two pieces of carpet $5\frac{1}{8}$ metres and $3\frac{3}{4}$ metres long are joined together. What is their total length?

43 $2\frac{2}{3}$ litres and $3\frac{5}{6}$ litres of oil are poured into a tin. How much oil will be in the tin?

44 I want a piece of wood $4\frac{3}{4}$ metres long. If I need to cut $\frac{5}{8}$ metres off a plank, how long is the plank?

45 Mr Heap's house stands on $2\frac{2}{3}$ hectares of land. If he buys another $\frac{4}{9}$ hectares, how much land will he have altogether?

46 The distance all the way round any shape is called its *perimeter*. Find the perimeters of these triangles.

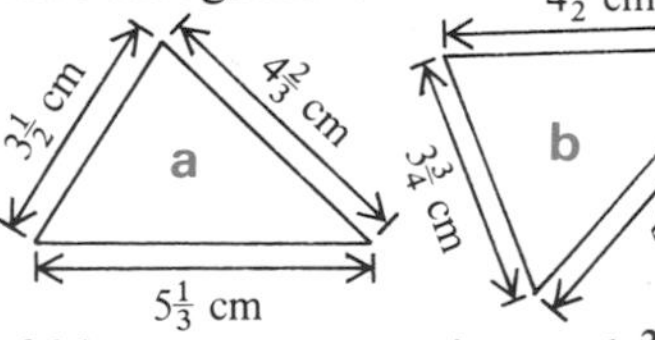

47 Jimmy Higgins spent $\frac{3}{4}$ of his money yesterday and $\frac{2}{9}$ of his money today. What fraction of his money has he **a** spent **b** left?

48 Mrs Armstrong carries home three bags of shopping with masses of $5\frac{1}{3}$ kg, $8\frac{3}{4}$ kg and $6\frac{5}{6}$ kg. What is the total mass of her shopping?

49 Mr Garforth spent $\frac{1}{3}$ of his life as a fisherman, $\frac{1}{8}$ of his life as a market trader and $\frac{1}{6}$ of his life as a warehouseman. What fraction of his life was spent working?

50 Find the perimeters of these three rectangles.

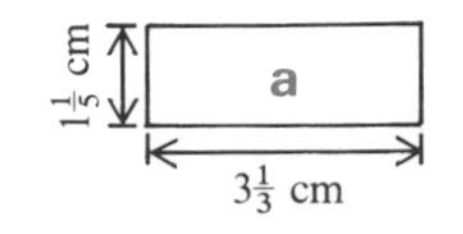

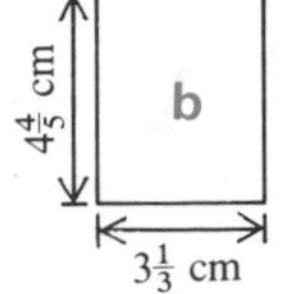

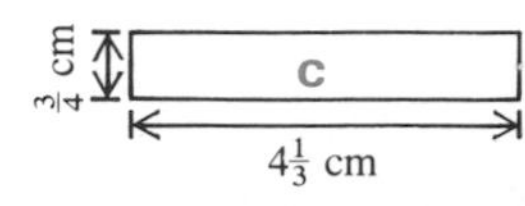

Subtraction of fractions

Part 1

Work the following.

1 a $1 - \frac{1}{4}$ b $1 - \frac{1}{5}$ c $1 - \frac{1}{6}$ d $1 - \frac{2}{3}$
e $1 - \frac{4}{9}$ f $1 - \frac{5}{6}$ g $1 - \frac{3}{5}$ h $1 - \frac{9}{10}$
i $1 - \frac{5}{8}$ j $1 - \frac{7}{8}$

2 a $2 - \frac{1}{3}$ b $3 - \frac{3}{4}$ c $4 - \frac{2}{5}$ d $9 - \frac{3}{8}$
e $3 - \frac{9}{10}$ f $4 - \frac{7}{10}$ g $3 - \frac{11}{12}$ h $5 - \frac{7}{8}$
i $10 - \frac{7}{12}$ j $12 - \frac{3}{20}$

3 a $4 - 1\frac{1}{2}$ b $5 - 2\frac{1}{4}$ c $6 - 1\frac{1}{3}$ d $8 - 3\frac{3}{4}$
e $7 - 2\frac{9}{10}$ f $6 - 1\frac{7}{8}$ g $8 - 2\frac{3}{5}$ h $7 - 1\frac{5}{8}$
i $6 - 4\frac{3}{10}$ j $4 - 3\frac{3}{4}$

4 a $5\frac{2}{3} - 3\frac{1}{3}$ b $4\frac{3}{4} - 1\frac{1}{4}$ c $2\frac{4}{5} - 1\frac{3}{5}$ d $8\frac{7}{9} - 2\frac{5}{9}$
e $13\frac{2}{7} - 3\frac{1}{7}$ f $6\frac{3}{5} - 2\frac{1}{5}$ g $9\frac{2}{3} - 8\frac{1}{3}$ h $9\frac{4}{11} - 3\frac{2}{11}$
i $8\frac{4}{5} - 7\frac{1}{5}$ j $4\frac{6}{7} - 3\frac{2}{7}$

5 a $3\frac{1}{3} - \frac{2}{3}$ b $4\frac{2}{5} - \frac{3}{5}$ c $8\frac{3}{7} - \frac{5}{7}$ d $5\frac{1}{5} - \frac{4}{5}$
e $6\frac{2}{9} - \frac{7}{9}$ f $4\frac{3}{10} - \frac{7}{10}$ g $7\frac{1}{5} - \frac{4}{5}$ h $8\frac{2}{7} - \frac{5}{7}$
i $1\frac{4}{11} - \frac{7}{11}$ j $2\frac{3}{10} - \frac{9}{10}$

6 a $2\frac{1}{3} - 1\frac{2}{3}$ b $3\frac{1}{4} - 1\frac{3}{4}$ c $5\frac{2}{5} - 3\frac{3}{5}$ d $7\frac{2}{5} - 1\frac{3}{5}$
e $4\frac{1}{4} - 3\frac{3}{4}$ f $7\frac{4}{7} - 3\frac{5}{7}$ g $8\frac{1}{3} - 6\frac{2}{3}$ h $2\frac{1}{9} - 1\frac{8}{9}$
i $3\frac{2}{7} - 2\frac{3}{7}$ j $8\frac{3}{10} - 7\frac{4}{10}$ k $4\frac{2}{9} - 1\frac{5}{9}$ l $8\frac{1}{3} - 1\frac{2}{3}$
m $12\frac{4}{9} - 7\frac{7}{9}$ n $9\frac{2}{5} - 8\frac{3}{5}$ o $7\frac{4}{6} - 3\frac{5}{6}$ p $8\frac{4}{10} - 7\frac{7}{10}$
q $5\frac{2}{5} - 4\frac{4}{5}$ r $7\frac{4}{9} - 6\frac{7}{9}$ s $9\frac{1}{5} - 7\frac{4}{5}$ t $3\frac{3}{13} - 2\frac{8}{13}$

Part 2

Copy and complete the following.

1 $\frac{3}{4} - \frac{1}{8} = \frac{}{8} - \frac{1}{8} = \frac{}{8}$ **2** $\frac{4}{5} - \frac{7}{10} = \frac{}{10} - \frac{7}{10} = \frac{}{10}$
3 $\frac{2}{3} - \frac{1}{12} = \frac{}{12} - \frac{1}{12} = \frac{}{12}$ **4** $\frac{3}{5} - \frac{3}{20} = \frac{}{20} - \frac{3}{20} = \frac{}{20}$
5 $\frac{1}{2} - \frac{3}{8} = \frac{}{8} - \frac{3}{8} = \frac{}{8}$ **6** $\frac{2}{5} - \frac{2}{15} = \frac{}{15} - \frac{2}{15} = \frac{}{15}$
7 $\frac{2}{3} - \frac{4}{9} = \frac{}{9} - \frac{4}{9} = \frac{}{9}$ **8** $\frac{7}{8} - \frac{1}{4} = \frac{7}{8} - \frac{}{8} = \frac{}{8}$
9 $\frac{11}{12} - \frac{1}{3} = \frac{11}{12} - \frac{}{12} = \frac{}{12}$ **10** $\frac{11}{18} - \frac{2}{9} = \frac{11}{18} - \frac{}{18} = \frac{}{18}$

Work the following.

11 $\frac{3}{4} - \frac{3}{8}$ **12** $\frac{2}{3} - \frac{7}{12}$ **13** $\frac{3}{5} - \frac{3}{10}$ **14** $\frac{6}{7} - \frac{2}{21}$
15 $\frac{4}{5} - \frac{3}{20}$ **16** $\frac{23}{24} - \frac{2}{3}$ **17** $\frac{17}{18} - \frac{1}{3}$ **18** $\frac{9}{10} - \frac{3}{5}$

Subtraction of fractions

Cancel the answers where possible.

19 $\frac{7}{12} - \frac{1}{3}$ **20** $\frac{14}{20} - \frac{1}{2}$ **21** $\frac{13}{15} - \frac{2}{3}$ **22** $\frac{11}{18} - \frac{1}{2}$

23 $\frac{11}{24} - \frac{1}{3}$ **24** $\frac{3}{4} - \frac{5}{12}$ **25** $\frac{5}{6} - \frac{7}{12}$ **26** $\frac{7}{8} - \frac{5}{24}$

27 $\frac{3}{4} - \frac{5}{28}$ **28** $\frac{7}{10} - \frac{9}{20}$ **29** $\frac{2}{3} - \frac{1}{6}$ **30** $\frac{5}{8} - \frac{11}{24}$

31 $3\frac{11}{12} - \frac{3}{4}$ **32** $4\frac{13}{18} - \frac{5}{9}$ **33** $8\frac{11}{18} - \frac{1}{9}$ **34** $5\frac{1}{4} - \frac{1}{12}$

35 $7\frac{7}{8} - \frac{1}{2}$ **36** $2\frac{7}{10} - \frac{2}{5}$ **37** $1\frac{8}{15} - \frac{2}{5}$ **38** $4\frac{1}{2} - \frac{7}{18}$

39 $8\frac{23}{24} - \frac{2}{3}$ **40** $6\frac{7}{8} - \frac{3}{4}$ **41** $7\frac{9}{10} - \frac{2}{5}$ **42** $8\frac{11}{12} - \frac{1}{2}$

43 $5\frac{1}{2} - 3\frac{1}{12}$ **44** $9\frac{13}{21} - 2\frac{1}{3}$ **45** $4\frac{7}{8} - 1\frac{1}{2}$ **46** $3\frac{8}{9} - 2\frac{2}{3}$

47 $7\frac{2}{3} - 5\frac{5}{12}$ **48** $6\frac{1}{2} - 2\frac{3}{10}$ **49** $9\frac{3}{4} - 1\frac{5}{8}$ **50** $8\frac{9}{10} - 3\frac{2}{5}$

51 $6\frac{5}{6} - 2\frac{7}{18}$ **52** $4\frac{11}{16} - 2\frac{1}{2}$ **53** $7\frac{2}{3} - 5\frac{2}{9}$ **54** $3\frac{19}{24} - 1\frac{1}{6}$

The following need particular care.

55 $5\frac{1}{2} - 2\frac{3}{4}$ **56** $8\frac{1}{2} - 4\frac{3}{4}$ **57** $6\frac{1}{4} - 3\frac{1}{2}$ **58** $9\frac{1}{4} - 7\frac{1}{2}$

59 $8\frac{1}{8} - 3\frac{3}{4}$ **60** $3\frac{3}{8} - 1\frac{3}{4}$ **61** $7\frac{1}{8} - 2\frac{1}{2}$ **62** $5\frac{5}{12} - 2\frac{3}{4}$

63 $8\frac{7}{12} - 3\frac{5}{6}$ **64** $6\frac{1}{2} - 4\frac{7}{12}$ **65** $8\frac{1}{2} - 4\frac{7}{10}$ **66** $3\frac{1}{4} - 1\frac{7}{12}$

67 $5\frac{5}{9} - 2\frac{2}{3}$ **68** $4\frac{1}{9} - 2\frac{1}{3}$ **69** $7\frac{1}{5} - 3\frac{3}{10}$ **70** $4\frac{5}{12} - 3\frac{2}{3}$

Part 3

Copy and complete the following.

1 $\frac{3}{4} - \frac{2}{3} = \frac{}{12} - \frac{}{12} = \frac{}{12}$ **2** $\frac{2}{3} - \frac{1}{2} = \frac{}{6} - \frac{}{6} = \frac{}{6}$

3 $\frac{4}{5} - \frac{1}{2} = \frac{}{10} - \frac{}{10} = \frac{}{10}$ **4** $\frac{2}{3} - \frac{3}{5} = \frac{}{15} - \frac{}{15} = \frac{}{15}$

5 $\frac{1}{3} - \frac{1}{4} = \frac{}{12} - \frac{}{12} = \frac{}{12}$ **6** $\frac{1}{2} - \frac{2}{5} = \frac{}{10} - \frac{}{10} = \frac{}{10}$

7 $\frac{2}{3} - \frac{5}{8} = \frac{}{24} - \frac{}{24} = \frac{}{24}$ **8** $\frac{4}{5} - \frac{1}{3} = \frac{}{15} - \frac{}{15} = \frac{}{15}$

9 $\frac{3}{4} - \frac{1}{6} = \frac{}{12} - \frac{}{12} = \frac{}{12}$ **10** $\frac{5}{6} - \frac{3}{8} = \frac{}{24} - \frac{}{24} = \frac{}{24}$

Work these subtractions.

11 $\frac{1}{2} - \frac{2}{7}$ **12** $\frac{3}{4} - \frac{2}{5}$ **13** $\frac{2}{3} - \frac{1}{4}$ **14** $\frac{7}{8} - \frac{1}{3}$

15 $\frac{4}{5} - \frac{1}{4}$ **16** $\frac{5}{8} - \frac{2}{5}$ **17** $\frac{3}{5} - \frac{1}{2}$ **18** $\frac{7}{9} - \frac{1}{2}$

19 $8\frac{2}{3} - 4\frac{2}{5}$ **20** $7\frac{3}{4} - 4\frac{3}{5}$ **21** $4\frac{7}{8} - 2\frac{2}{3}$ **22** $6\frac{4}{5} - 3\frac{2}{9}$

23 $3\frac{7}{10} - 2\frac{1}{3}$ **24** $9\frac{5}{8} - 3\frac{2}{9}$ **25** $7\frac{7}{9} - 3\frac{1}{4}$ **26** $4\frac{9}{10} - 3\frac{3}{4}$

27 $8\frac{4}{7} - \frac{2}{5}$ **28** $2\frac{7}{9} - \frac{3}{4}$ **29** $9\frac{3}{4} - 2\frac{1}{10}$ **30** $3\frac{7}{12} - 1\frac{2}{9}$

31 $6\frac{13}{16} - 4\frac{2}{3}$ **32** $9\frac{4}{5} - 1\frac{3}{8}$ **33** $8\frac{11}{12} - 2\frac{4}{9}$ **34** $5\frac{9}{10} - 3\frac{1}{6}$

35 $3\frac{17}{20} - 2\frac{2}{3}$ **36** $5\frac{4}{5} - 1\frac{5}{12}$ **37** $7\frac{5}{6} - 3\frac{3}{10}$ **38** $8\frac{7}{8} - 7\frac{5}{6}$

Subtraction of fractions

The following need particular care.

39 $1\frac{1}{3} - \frac{3}{4}$	**40** $3\frac{1}{4} - \frac{2}{3}$	**41** $2\frac{2}{5} - \frac{3}{4}$	**42** $4\frac{1}{4} - \frac{4}{5}$
43 $7\frac{2}{5} - 3\frac{1}{2}$	**44** $6\frac{1}{3} - 4\frac{1}{2}$	**45** $5\frac{1}{4} - 2\frac{3}{5}$	**46** $9\frac{3}{8} - 4\frac{2}{3}$
47 $8\frac{5}{12} - 2\frac{7}{9}$	**48** $4\frac{1}{5} - 1\frac{4}{9}$	**49** $2\frac{3}{10} - 1\frac{5}{6}$	**50** $8\frac{3}{8} - 4\frac{5}{6}$
51 $3\frac{3}{16} - 2\frac{2}{3}$	**52** $5\frac{1}{6} - 4\frac{9}{10}$	**53** $6\frac{2}{7} - 2\frac{4}{5}$	**54** $7\frac{2}{9} - 6\frac{11}{12}$

Part 4

1. If I save $\frac{3}{5}$ of what I earn, what fraction of my earnings do I spend?
2. $\frac{5}{8}$ of my working day is spent in my office. What fraction of the day is spent elsewhere?
3. I post a large box with a total mass of $5\frac{7}{8}$ kg. If the box and wrappings have a mass of $1\frac{1}{4}$ kg, what is the mass of the contents?
4. My living room is $8\frac{5}{8}$ metres long, and I buy a carpet with a length of $7\frac{1}{2}$ metres. What is the length of the room not covered by the carpet?
5. I have a $7\frac{1}{2}$-litre tin of oil from which I pour $4\frac{3}{8}$ litres. How much oil is left in the tin?
6. A plank of wood measures $6\frac{3}{8}$ metres and I cut off $4\frac{1}{2}$ metres. What length of wood is left?
7. Mr Green has $1\frac{3}{5}$ hectares of land around his house and he plants vegetables on $\frac{3}{8}$ hectare. What area of land does he use for other purposes?
8. If AC = $9\frac{1}{6}$ cm and AB = $5\frac{2}{3}$ cm, find the length of BC.

9. $\frac{1}{4}$ of my salary is spent on my mortgage and $\frac{2}{3}$ is spent on food and fuel. What fraction is left for other things?
10. Farmer Giles has $\frac{1}{8}$ of his land as an orchard, $\frac{1}{2}$ of his land to grass, and $\frac{1}{3}$ under the plough. What fraction of his land is put to other uses?

Fractions of quantities

Part 1

Find the following fractions.

1 $\frac{2}{3}$ of £3·96
2 $\frac{3}{4}$ of £8·84
3 $\frac{2}{5}$ of £6·55
4 $\frac{3}{8}$ of £25·68
5 $\frac{4}{5}$ of £17·10
6 $\frac{5}{6}$ of £27·36
7 $\frac{2}{3}$ of £22·95
8 $\frac{4}{9}$ of £38·25
9 $\frac{3}{10}$ of £15·90
10 $\frac{5}{7}$ of £52·71

11 Mr Dunhill earns £65·25 in one week. After tax is deducted, he receives only two thirds of this amount. How much does he receive?

12 Susan gets £1·35 pocket money each week. She saves two fifths of it and spends the rest. Find **a** how much she saves **b** how much she spends.

13 A bag of flour has a mass of 1.25 kg, but two fifths of it was lost through a hole in the side. What mass was lost?

14 Mr McNab had a mass of 62.3 kg before he went on a diet. By dieting, he lost two sevenths of this mass.
a How much did he lose? **b** What is his mass now?

15 A full lorry has a mass of 13.2 tonnes. Five eighths of this mass is the load and the rest is the mass of the lorry.
Find **a** the mass of the load **b** the mass of the lorry.

16 A new car cost Mr Davidson £2465·50. After two years he wanted to trade it in for another, and the garage offered him $\frac{3}{5}$ of this amount. How much was their offer worth?

17 A roll of curtain material which is 24 metres long, is found to have a flaw running $\frac{3}{10}$ of the way along its length.
a What length is faulty? **b** What length is still good?

18 Three quarters of the way from Bristol to Derby is by motorway. If the total distance is 142 miles, how many miles are
a motorway **b** ordinary roads?

19 A firm makes armchairs each with a mass of 102 kg. Three eighths of each chair is made of wood and the rest is of plastic. For each chair, what mass is used of
a wood **b** plastic?

20 A book has 850 pages. Three twentieths of the pages have pictures on them, and the rest are worded. How many pages have on them
a pictures **b** words?

Part 2

Find the following fractions.

1 $\frac{5}{6}$ of £7·38
2 $\frac{3}{8}$ of £19·68
3 $\frac{2}{5}$ of 12.4 metres
4 $\frac{3}{4}$ of 14.6 metres
5 $\frac{7}{10}$ of 49.2 kg
6 $\frac{4}{5}$ of 4.63 kg
7 $\frac{5}{12}$ of 28.2 litres
8 $\frac{17}{20}$ of 87.0 litres
9 $\frac{7}{11}$ of 478.5 km
10 $\frac{11}{12}$ of 726 km

11 A tank holds 2560 litres of petrol when full. It is only $\frac{7}{8}$ full. Find
a how many litres of petrol it has in it
b how many more litres it could take.

Fractions of quantities

12 $\frac{7}{12}$ of the total number of children in a school come by bus, $\frac{3}{10}$ walk the whole way and the rest of them come by car. If there are 1020 children in the school, find how many come **a** by bus **b** on foot **c** by car.

13 In one month a factory manufactures 36 450 vehicles. $\frac{8}{15}$ of these vehicles are cars, $\frac{4}{9}$ are lorries and the rest are coaches. Find how many of each type of vehicle they manufacture.

14 A bakery make 1350 loaves of bread on one night shift. $\frac{7}{10}$ are white, $\frac{4}{25}$ are brown and the rest are wheatmeal. Find how many of each type of loaf they have baked.

15 A glossy magazine has 60 pages of which $\frac{7}{12}$ are just for advertising. $\frac{4}{15}$ have photographs only on them and the rest are for articles and stories. How many pages have articles and stories on them?

16 A man takes 1 hour 15 minutes to get to work in the morning. $\frac{4}{5}$ of this time is spent on a bus and the bus is in heavy traffic for $\frac{1}{3}$ of the time he is on it. How long does the man spend on the bus in heavy traffic?

17 A farmer has 192 animals of which $\frac{7}{16}$ are cattle. Of the cattle, $\frac{2}{3}$ are dairy cows. How many dairy cows has he?

18 In a village there are 693 children. $\frac{4}{7}$ of the total are boys, and of these boys $\frac{8}{11}$ go to school. Of the boys at school, $\frac{2}{9}$ are at primary school. How many boys are there who go to primary school?

19 A company make 630 different kinds of toys, of which $\frac{4}{15}$ are designed for the under-5-year-olds. Of these toys for the under-fives, $\frac{2}{7}$ are sold overseas and Germany takes $\frac{1}{6}$ of the overseas sales of them. Find the number of types of toys for the under-fives which go to Germany.

20 $\frac{4}{9}$ of the houses on an estate are bungalows. $\frac{7}{12}$ of the bungalows have only two bedrooms. $\frac{8}{15}$ of the two-bedroomed bungalows have central heating. If there are 1620 houses on the estate, how many two-bedroomed bungalows do not have central heating?

Decimals

Preliminary Exercises

How many squares?

How many small squares are there in each of these diagrams?

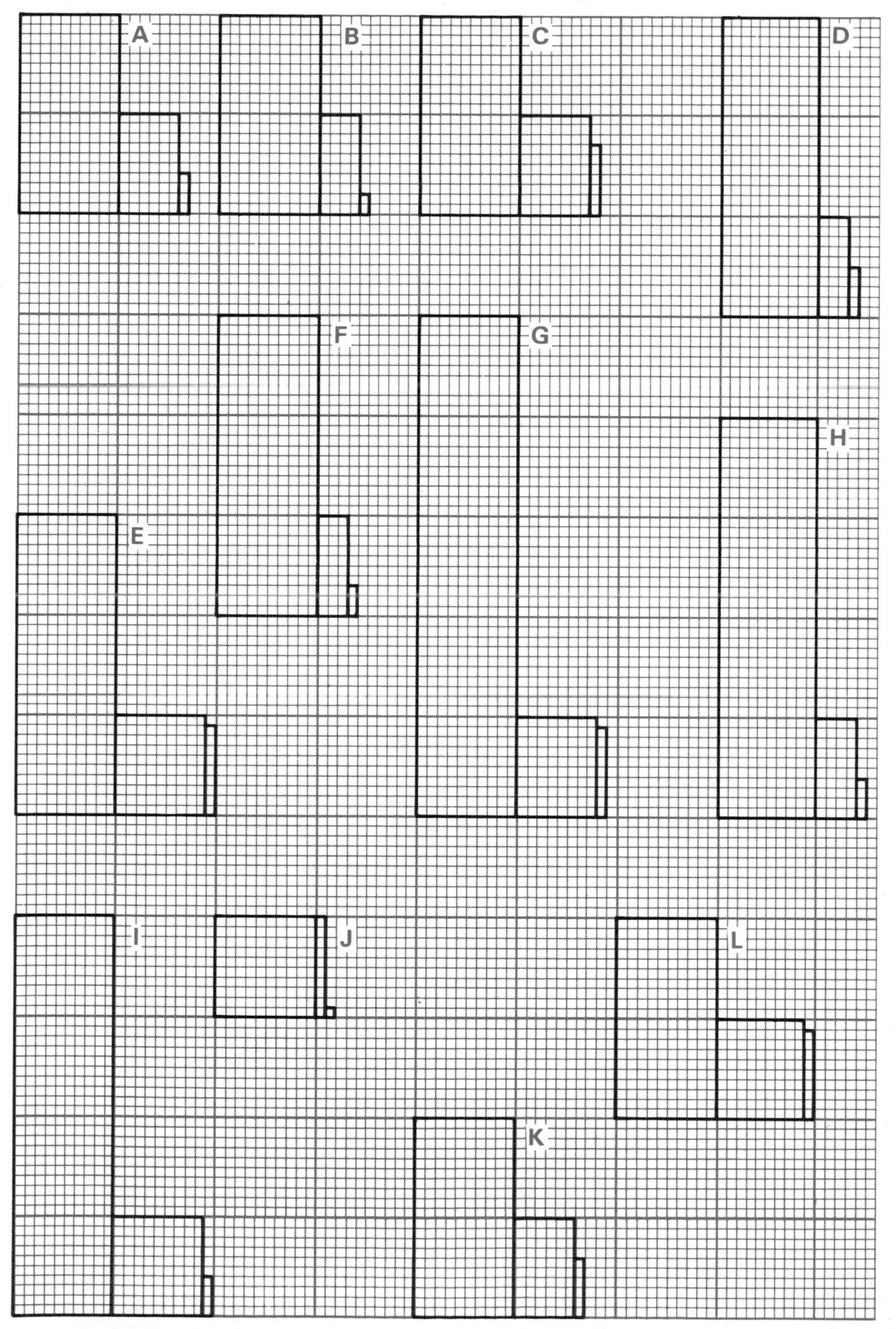

How many squares?

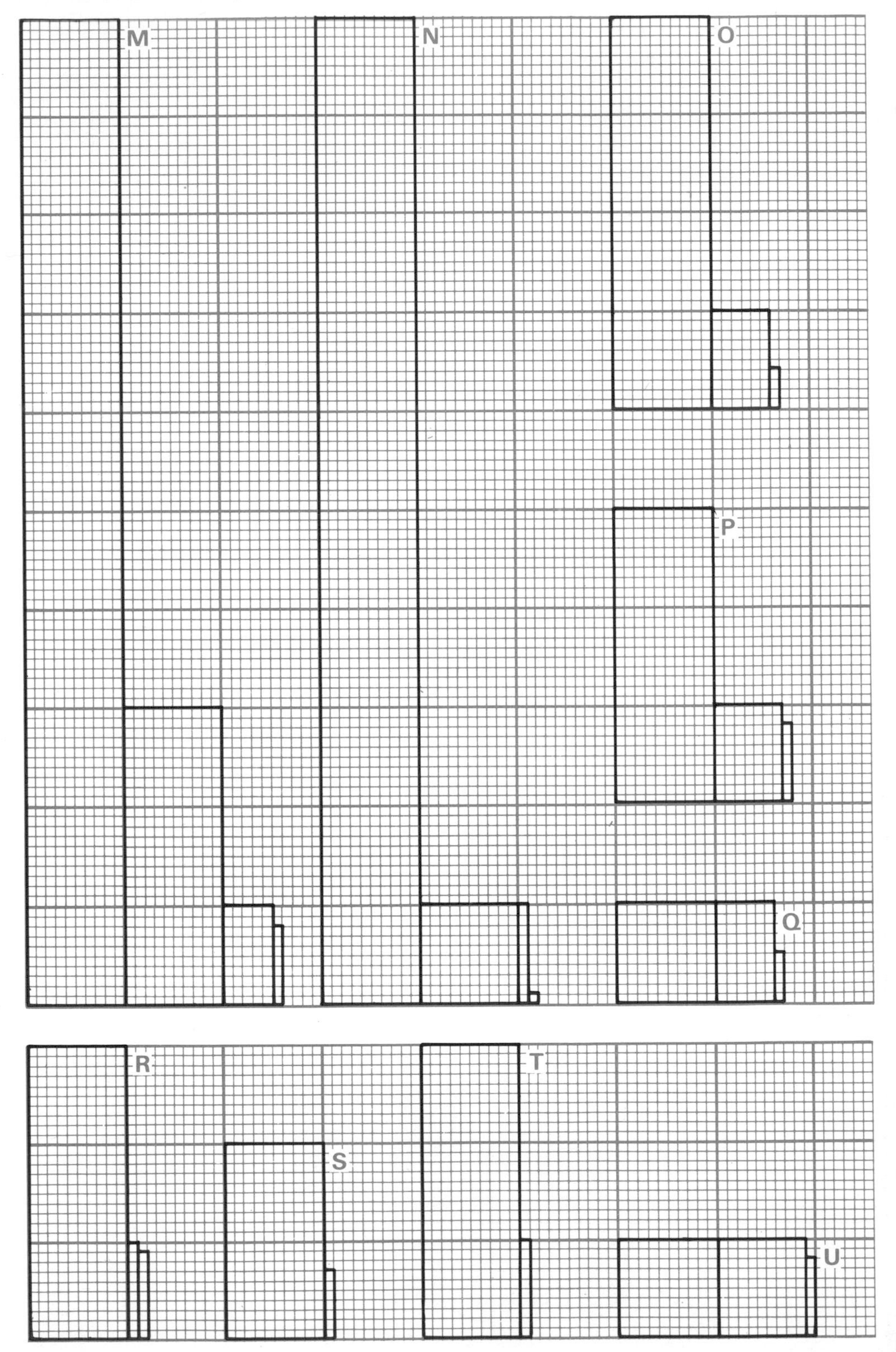

Numbers to size

If we draw the number 16.5
so that the height of each figure
is in proportion to its size,
we get a diagram as shown here.

The 1 is 10 cm high;
the 6 is 6 cm high;
the 5 is $\frac{5}{10}$ cm high.

Check this using a ruler.

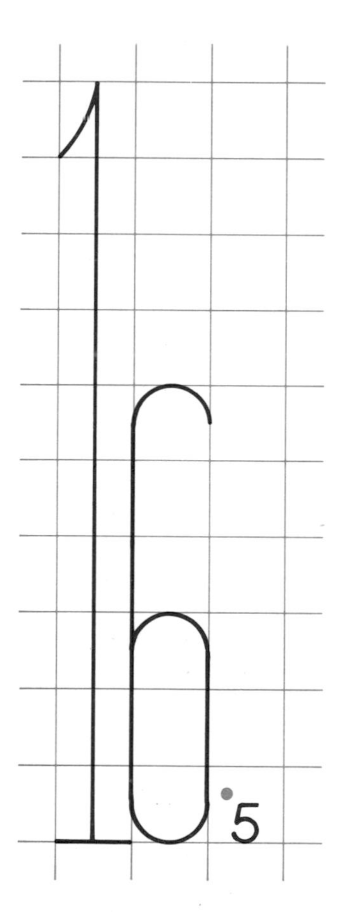

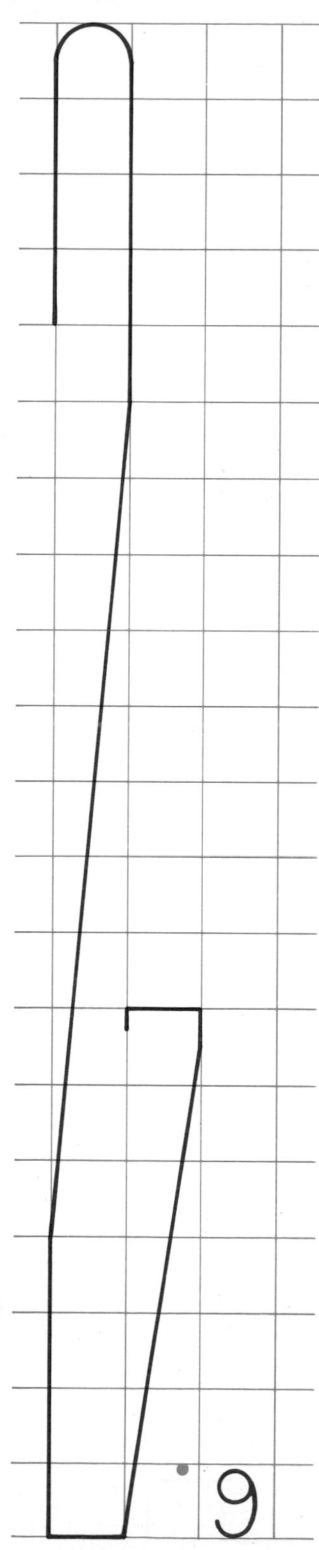

This example shows 27.9, drawn in the same way.

Numbers to size

A

Write the numbers which should be drawn into these blank diagrams.

Use a ruler to check your answers.

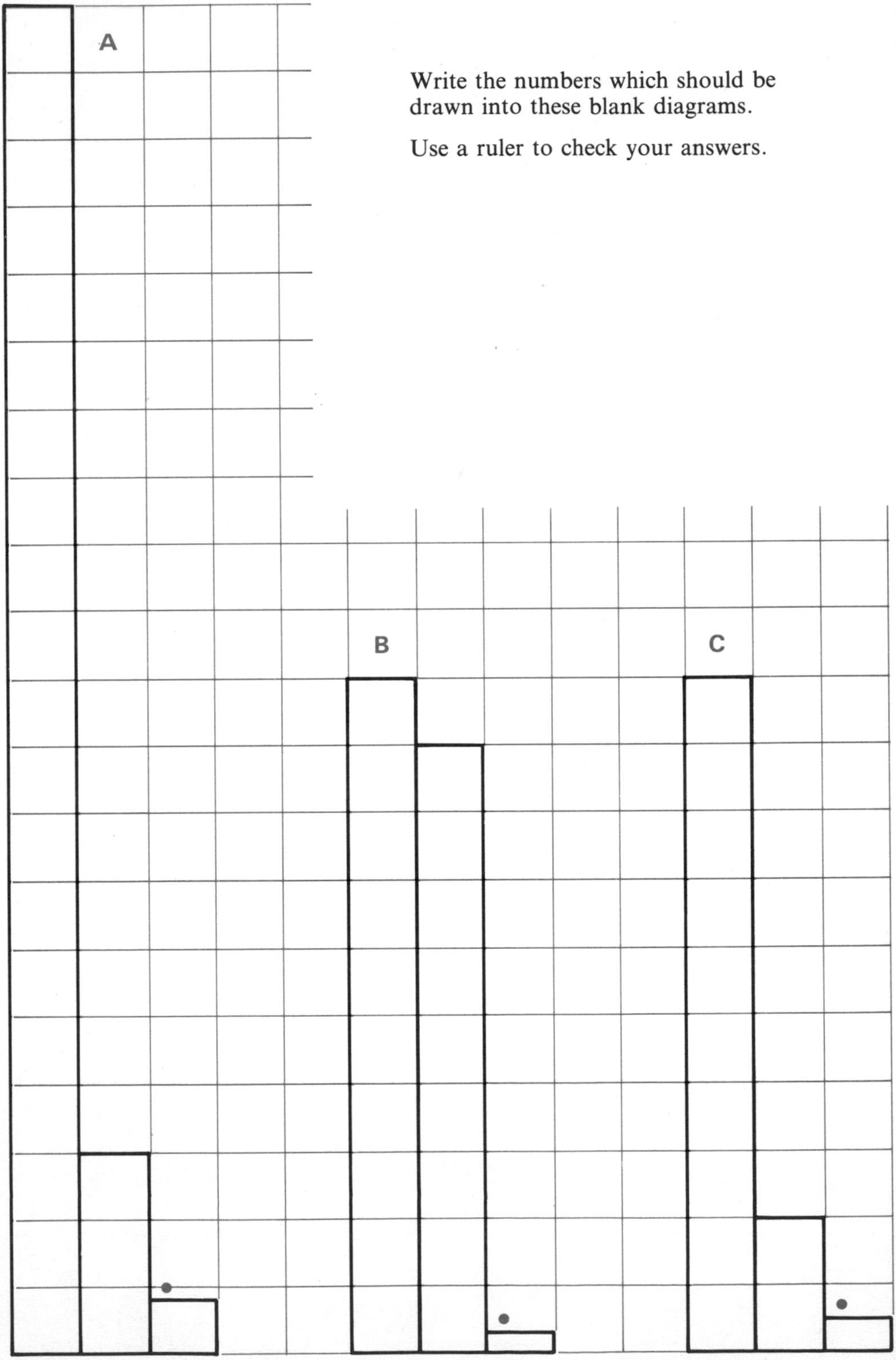

B

C

Numbers to size

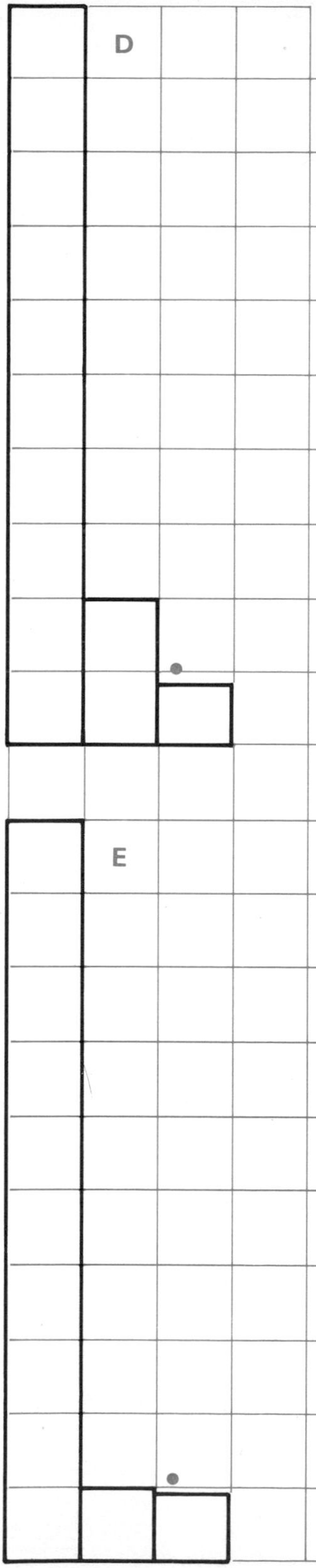

D

E

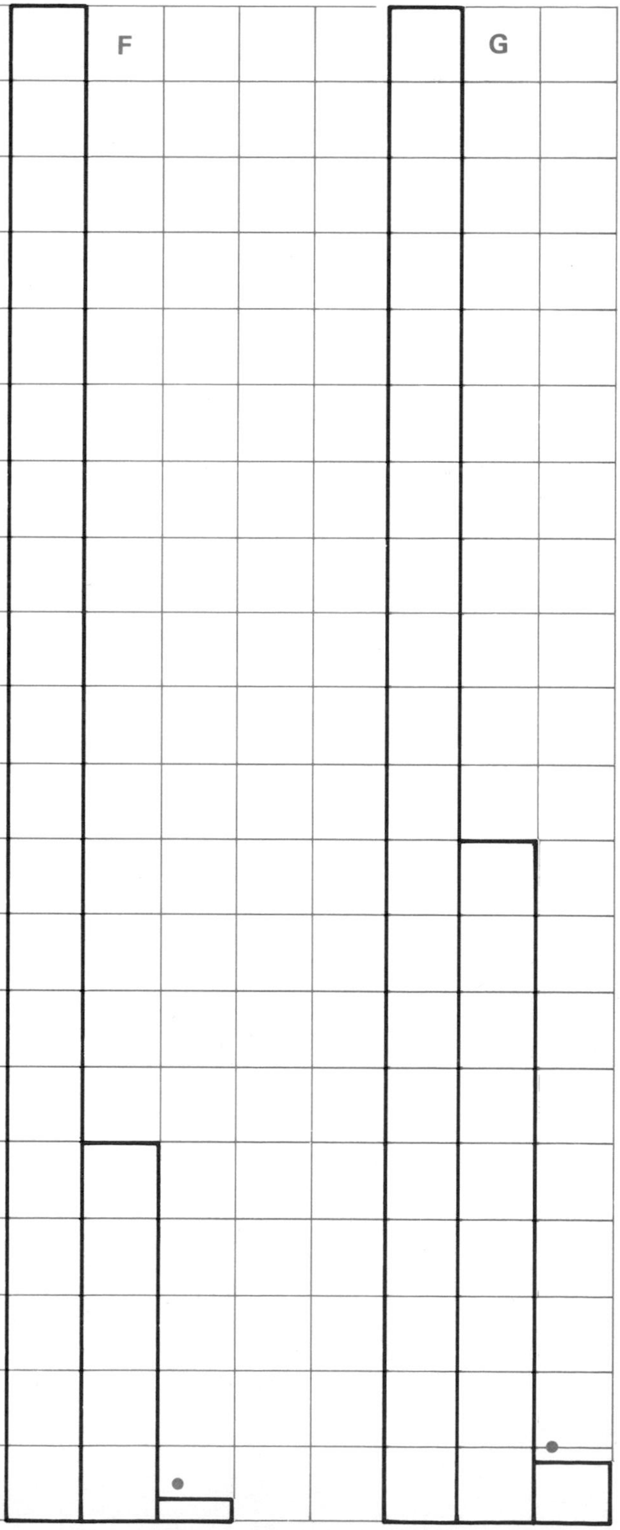

F

G

Numbers to size

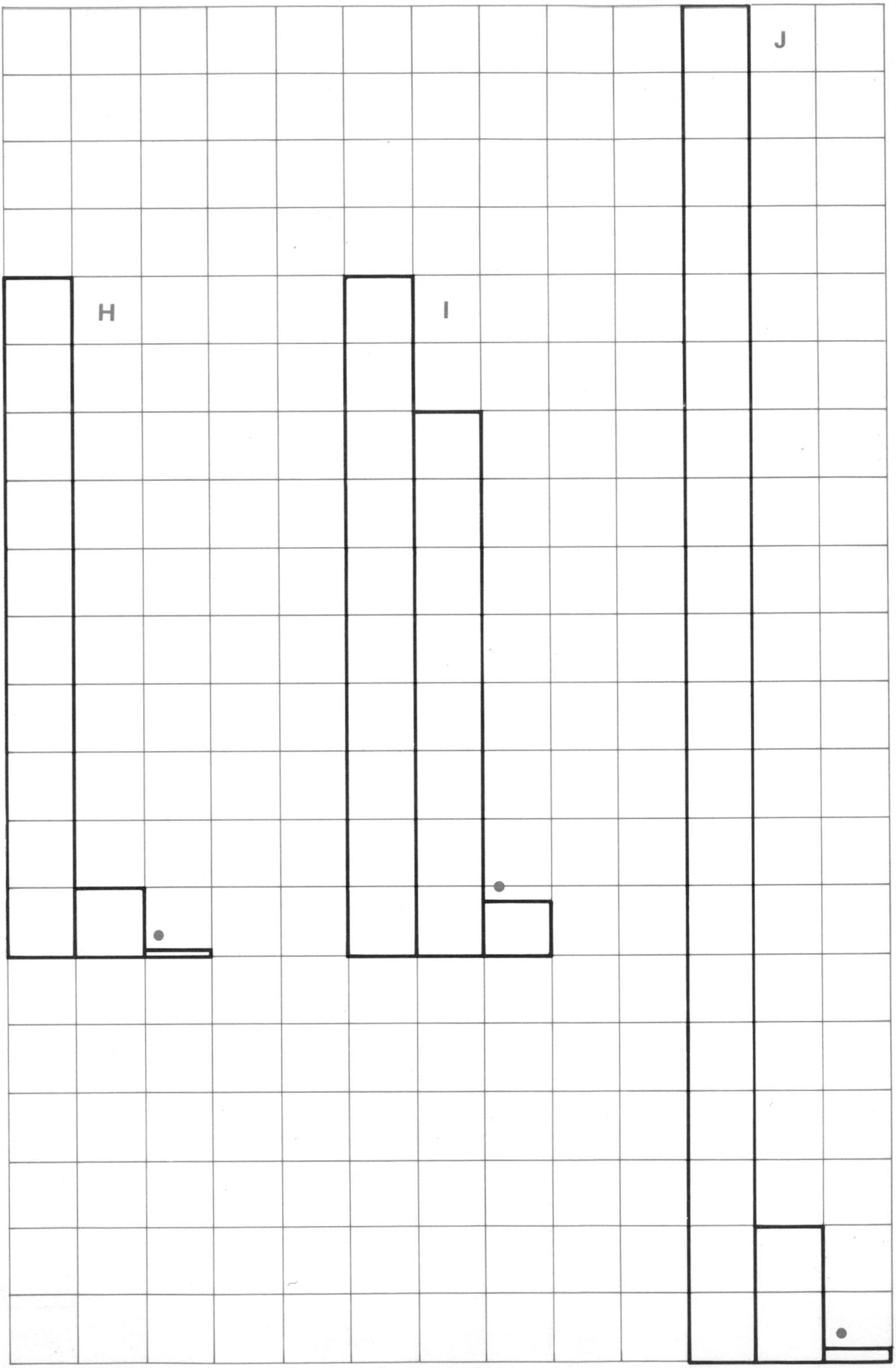

H

I

J

Measuring in decimals

1 Each of these jagged lines has three parts. Use your ruler to measure the **total** length of each line in **centimetres** and write down your results. You will need to use a decimal point.

2 Measure the lengths of the lines again, but this time measure in **millimetres**. Write down your results.

A B C D E F G H

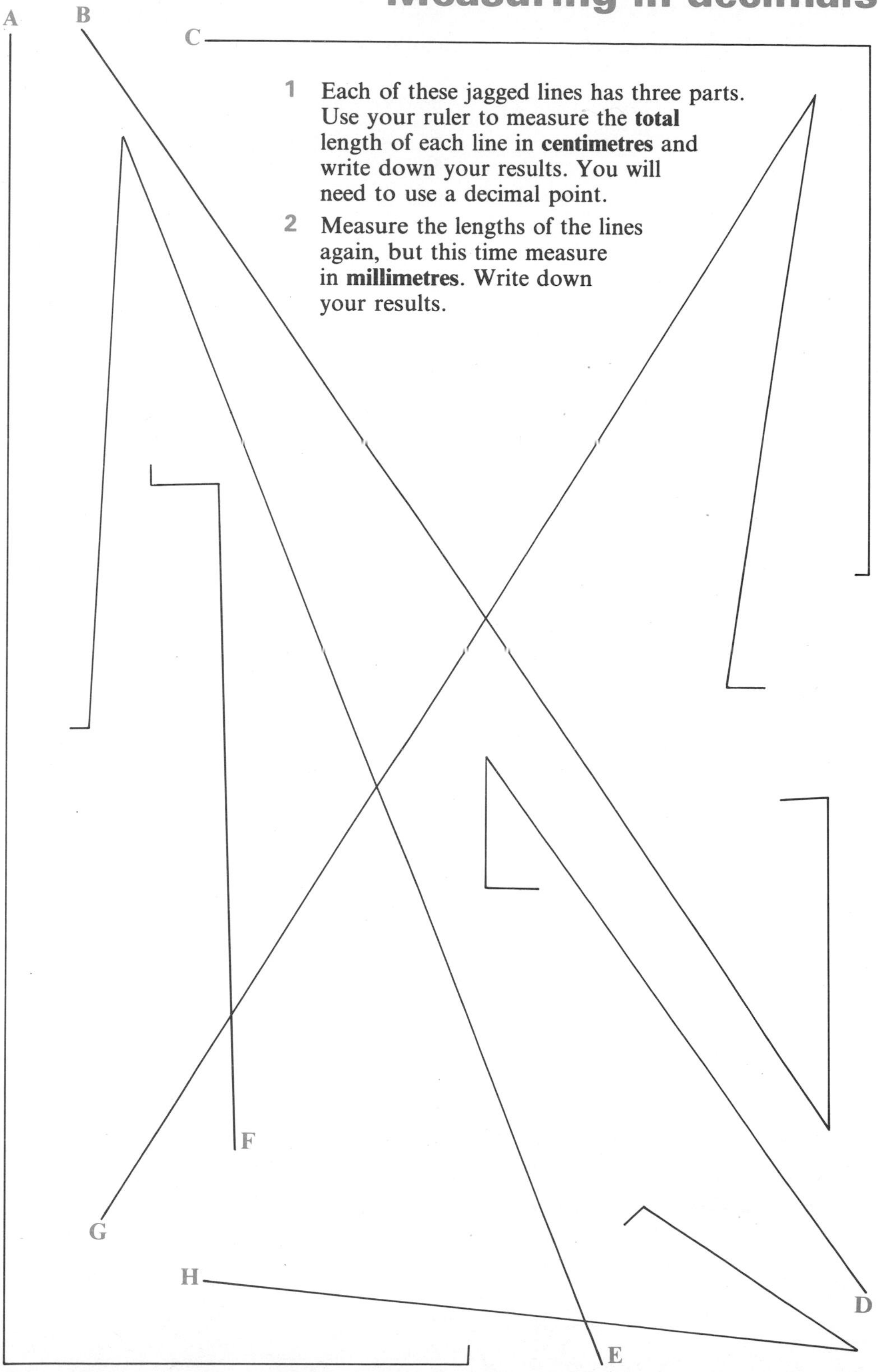

Measuring in decimals

3 Draw jagged lines of your own for the following lengths. Each line should have three parts.

a	23.6 cm	b	17.2 cm
c	12.7 cm	d	22.2 cm
e	19.1 cm	f	255 mm
g	176 mm	h	219 mm
i	198 mm	j	114 mm

4 a Measure the lengths of these jagged lines in centimetres. Write your results.
b Measure their lengths again in millimetres, and write your results.

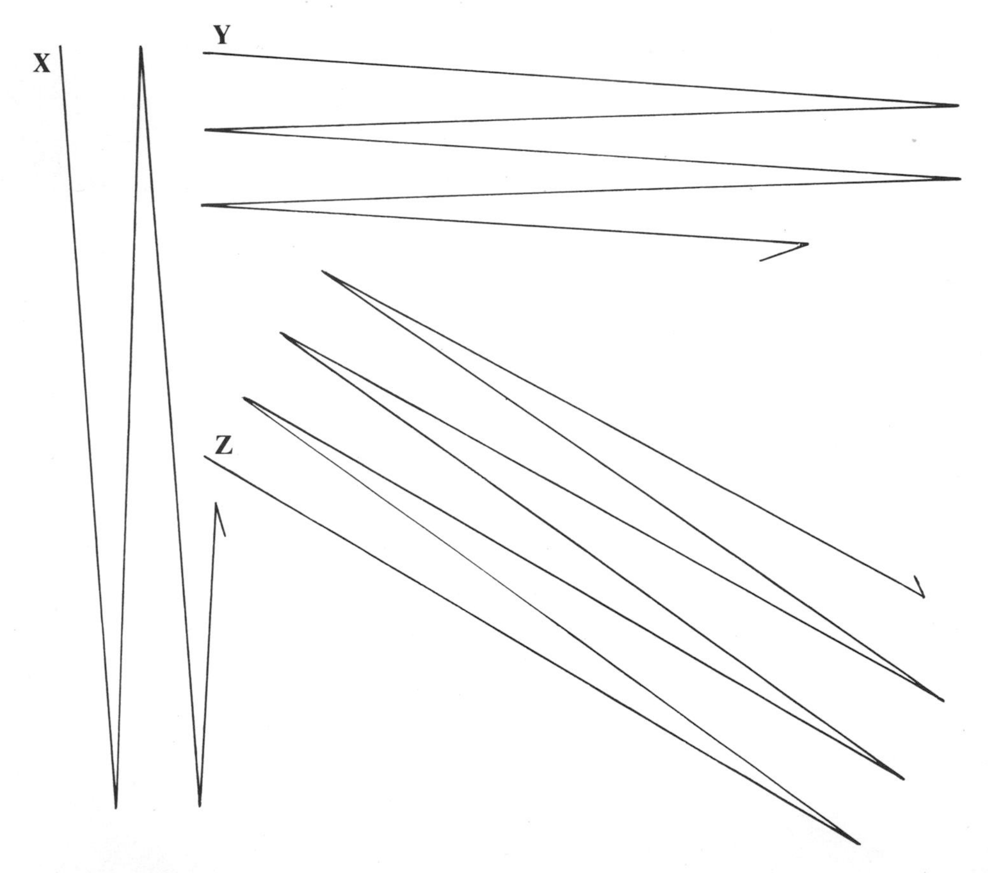

5 Draw jagged lines of your own for these lengths.

a	42.8 cm	b	37.9 cm	c	52.6 cm
d	262 mm	e	493 mm	f	704 mm

The decimal point

1 Write these numbers in a column using a decimal point. The first one is done for you.

	hundreds	tens	units	tenths	hundredths	as a decimal
Example	2	4	1	6	5	241.65
a	3	7	9	4	5	
b	4	2	3	7	3	
c	1	0	8	4	6	
d		2	1	5	7	
e		3	7	2		
f		6	0	1		
g			5	7		
h			3	9	4	
i	4	2	6	0	5	
j		7	8	3		
k		1	5			
l	3	6	2	0	6	
m			0	7	5	
n			0	0	8	

2 Write the decimals only in your exercise book.

	tenths	hundredths	thousandths	as a decimal
Example	1	4	2	0.142
a	3	1	6	
b	2	4	3	
c	1	0	4	
d	6	7		
e	5	4		
f	3	7		
g	8	4		
h		4	6	
i		5	1	
j		5		
k		2		
l			8	
m			7	
n	9	9	9	

The decimal point

3 Write the following numbers as decimals.

	hundredths	thousandths
a	6	7
b	2	5
c	8	1
d	7	
e	3	
f		6

	hundredths	thousandths
g		4
h		2
i	2	7
j	5	
k		3
l		9

4 Write these additions, using a decimal point.

Example $1 + \frac{2}{10} + \frac{6}{100} = 1.26$

a $3 + \frac{7}{10} + \frac{5}{100}$

b $8 + \frac{3}{10} + \frac{4}{100}$

c $5 + \frac{8}{10} + \frac{9}{100}$

d $9 + \frac{4}{10} + \frac{2}{100}$

e $7 + \frac{3}{100}$

f $6 + \frac{9}{100}$

g $40 + 6 + \frac{3}{10} + \frac{4}{100}$

h $30 + 9 + \frac{2}{10} + \frac{8}{100}$

i $60 + 7 + \frac{8}{100}$

j $200 + 70 + 6 + \frac{4}{100}$

k $500 + 8 + \frac{3}{10}$

l $900 + 7 + \frac{2}{10}$

m $4 + \frac{3}{10} + \frac{2}{100} + \frac{7}{1000}$

n $7 + \frac{1}{10} + \frac{8}{100} + \frac{6}{1000}$

o $50 + 7 + \frac{4}{100} + \frac{3}{1000}$

p $80 + 4 + \frac{6}{1000}$

q $600 + 50 + 1 + \frac{3}{100}$

r $200 + 50 + \frac{8}{10} + \frac{3}{1000}$

s $90 + 9 + \frac{2}{1000}$

t $600 + \frac{8}{1000}$

5 Write the following numbers, using a decimal point.

a eight and four-tenths

b seven and six-tenths

c five and two-tenths and seven-hundredths

d nine and eight-tenths and four-hundredths

e thirty seven and two-tenths and nine-hundredths

f sixty nine and four-tenths and six-hundredths

g fifty six and no-tenths and six-hundredths

h ninety eight and seven-hundredths

i nineteen and four-hundredths

j two and six-thousandths

k five and four-thousandths

l nine-hundredths

m two-hundredths and five-thousandths

n one-tenth and eight-thousandths

Pictures of decimals and fractions

1 Write:
(i) as a decimal and (ii) as a fraction that part of each square which is *shaded*.

a 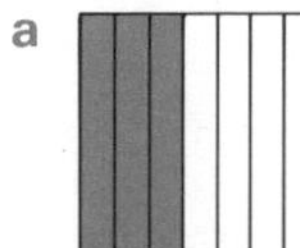 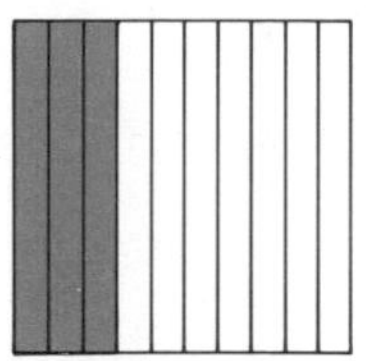

b 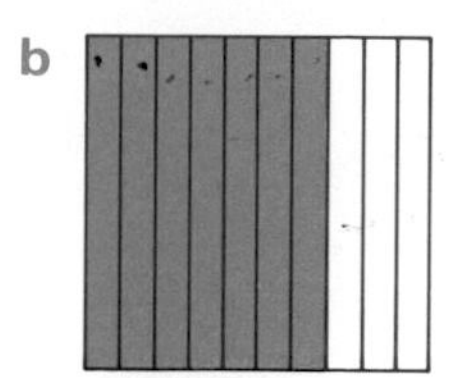

c 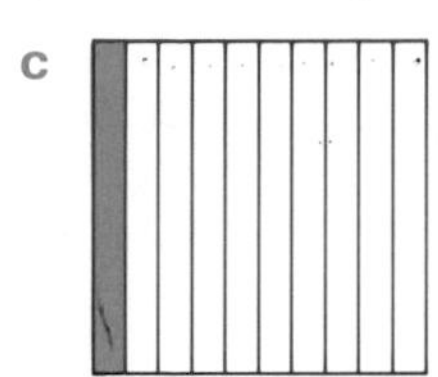

d 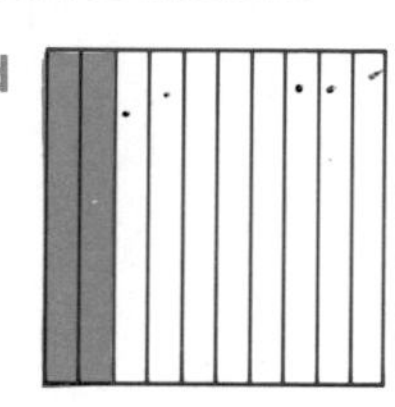

e 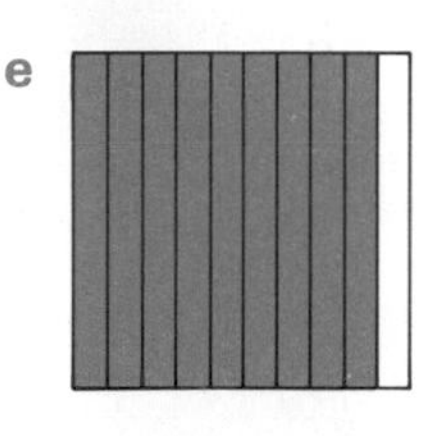

f 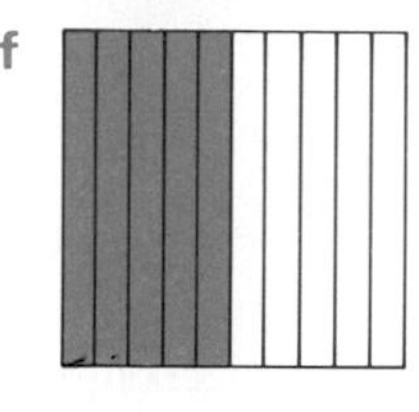

g 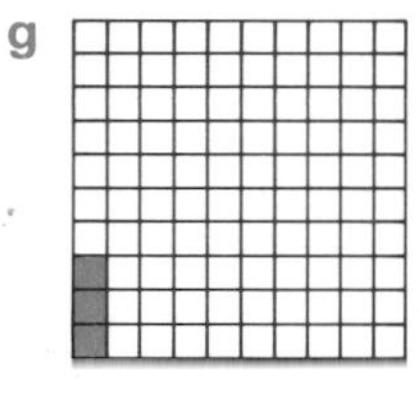

h 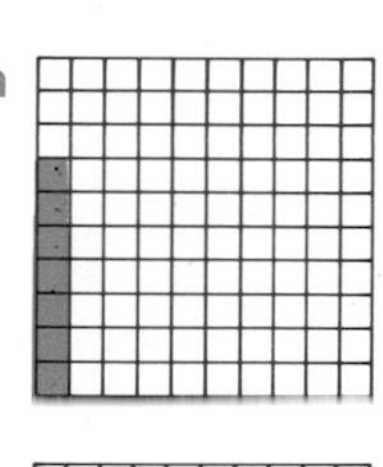

i 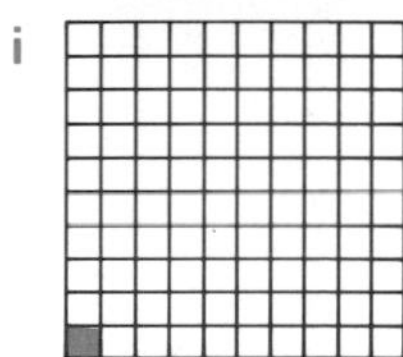

j 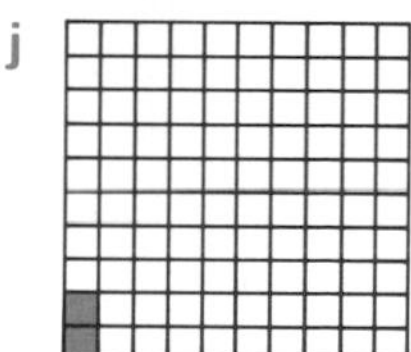

k 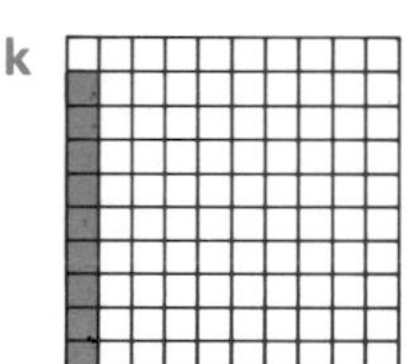

l 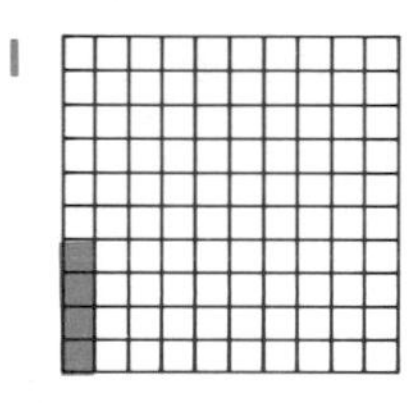

2 Write the fraction *shaded*, as in the example alongside.

Example

$\frac{2}{10} + \frac{3}{100} = \frac{23}{100} = 0.23$

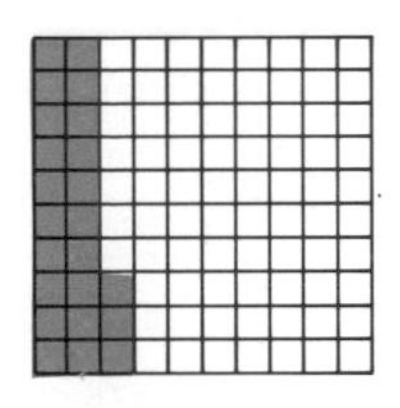

a 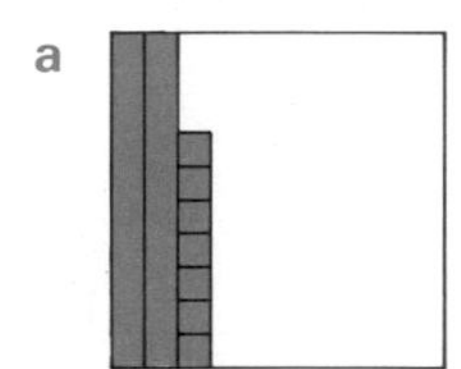

b 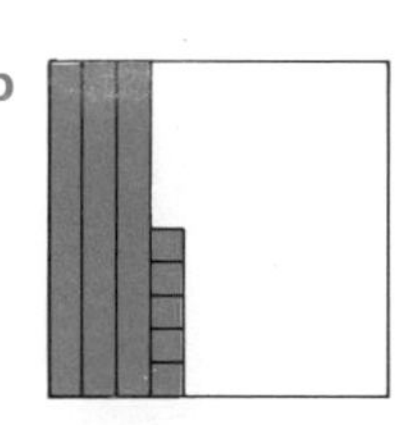

c 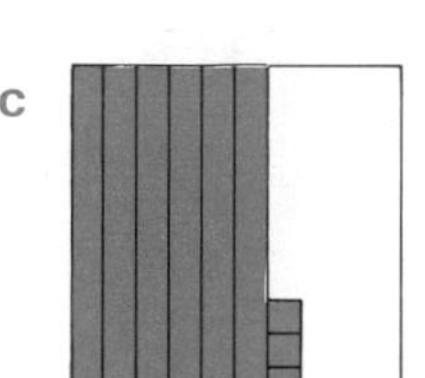

d 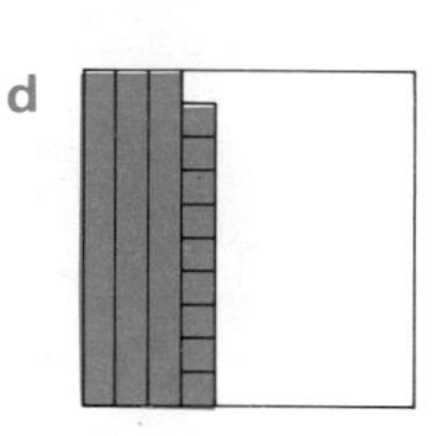

e 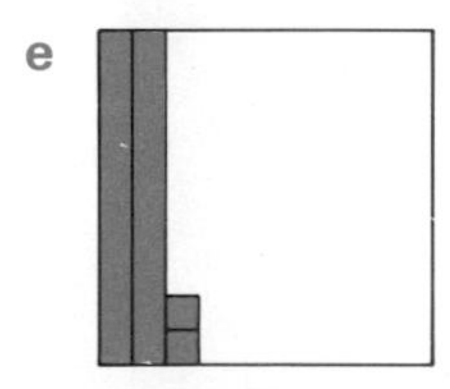

f 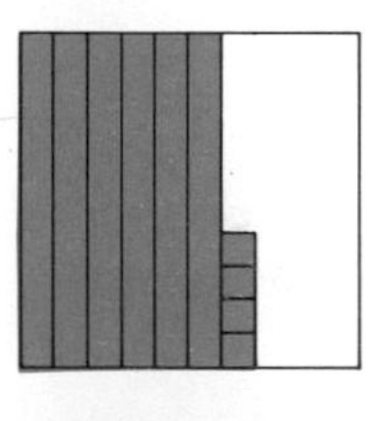

g 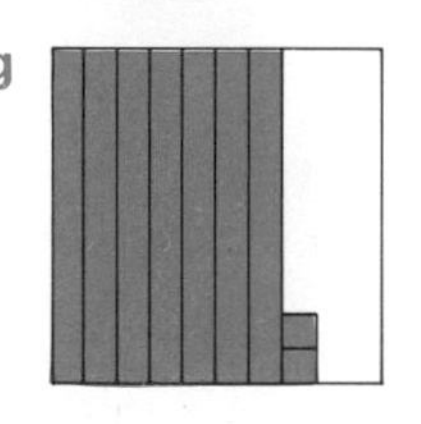

h 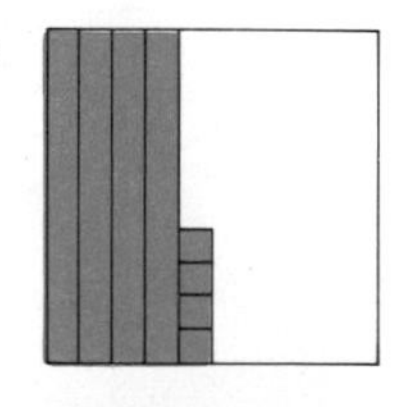

i 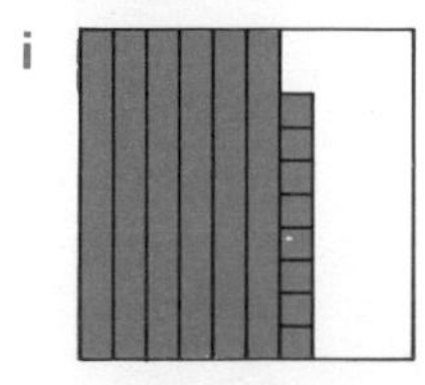

j 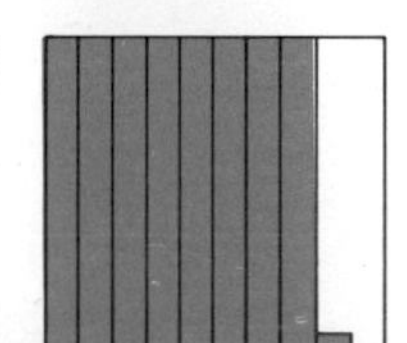

k 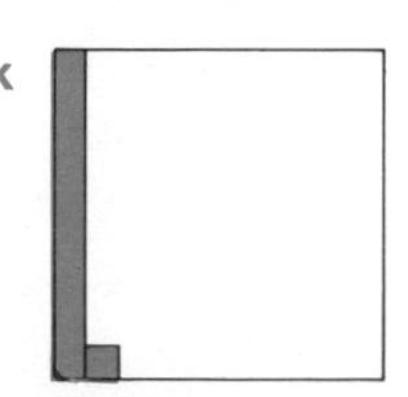

l 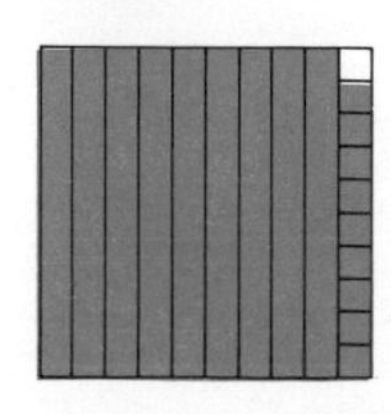

Pictures of decimals and fractions

3 Draw your own squares on to squared paper, and shade in the following fractions.

a $\frac{4}{10}$ **b** 0.6 **c** $\frac{4}{100}$ **d** 0.06 **e** 0.09
f 0.37 **g** 0.43 **h** $\frac{51}{100}$ **i** $\frac{7}{10}$ **j** $\frac{71}{100}$
k 0.73 **l** 0.8 **m** 0.94

4 Change these decimals to fractions.

a 0.2 **b** 0.6 **c** 0.7 **d** 0.9 **e** 0.1
f 0.01 **g** 0.08 **h** 0.03 **i** 0.24 **j** 0.74
k 0.13 **l** 0.99 **m** 0.001 **n** 0.007 **o** 0.009
p 0.079 **q** 0.033 **r** 0.057 **s** 0.243 **t** 0.624
u 0.999 **v** 0.13 **w** 0.27 **x** 0.5

5 Change these fractions to decimals.

a $\frac{3}{10}$ **b** $\frac{8}{10}$ **c** $\frac{1}{10}$ **d** $\frac{6}{10}$ **e** $\frac{3}{100}$
f $\frac{8}{100}$ **g** $\frac{1}{100}$ **h** $\frac{6}{100}$ **i** $\frac{3}{1000}$ **j** $\frac{8}{1000}$
k $\frac{1}{1000}$ **l** $\frac{6}{1000}$ **m** $\frac{16}{100}$ **n** $\frac{23}{100}$ **o** $\frac{87}{100}$
p $\frac{43}{100}$ **q** $\frac{245}{1000}$ **r** $\frac{625}{1000}$ **s** $\frac{132}{1000}$ **t** $\frac{67}{1000}$
u $\frac{28}{1000}$ **v** $\frac{49}{1000}$ **w** $\frac{7}{10}$ **x** $\frac{7}{100}$

6 Write each answer as a *decimal*.

a $\frac{3}{10} + \frac{4}{100}$ **b** $\frac{6}{10} + \frac{7}{100}$ **c** $\frac{1}{10} + \frac{8}{100}$
d $\frac{8}{10} + \frac{9}{100}$ **e** $\frac{6}{10} + \frac{4}{100} + \frac{3}{1000}$ **f** $\frac{5}{10} + \frac{2}{100} + \frac{1}{1000}$
g $\frac{4}{10} + \frac{3}{100} + \frac{9}{1000}$ **h** $\frac{9}{10} + \frac{0}{100} + \frac{4}{1000}$ **i** $\frac{6}{10} + \frac{4}{1000}$
j $\frac{8}{10} + \frac{6}{100} + \frac{7}{1000}$ **k** $\frac{0}{10} + \frac{4}{100} + \frac{2}{1000}$ **l** $\frac{5}{100} + \frac{9}{1000}$
m $\frac{2}{10} + \frac{1}{1000}$ **n** $\frac{3}{100} + \frac{4}{1000}$ **o** $\frac{2}{10} + \frac{5}{1000}$

Reading a scale

Write the number to which each lettered arrow is pointing.
Not every number will need a decimal point.

Example In **1**, A = 1.3

1

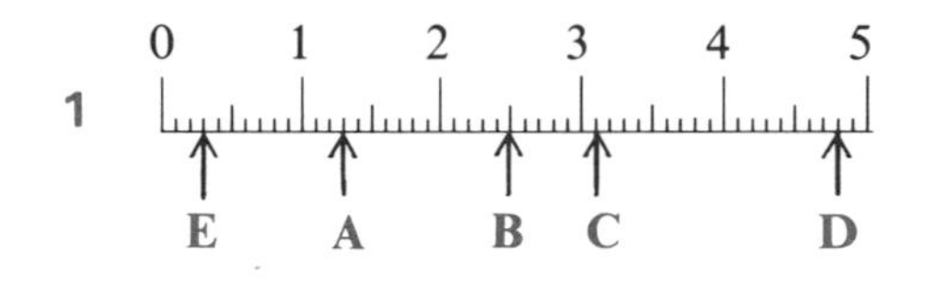

50 51 52 53 54 55

F G H I

6 7 8 9

J K L

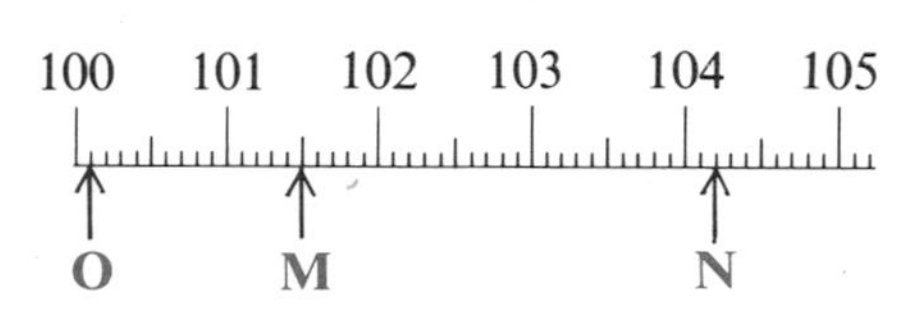

20 30 40 50 60

P Q R

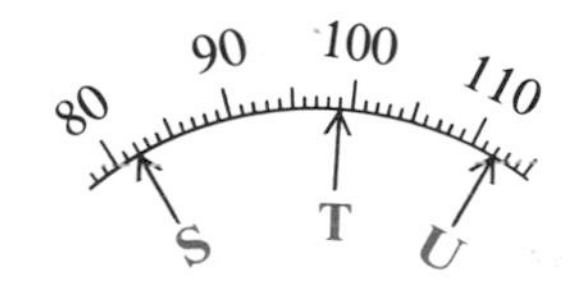

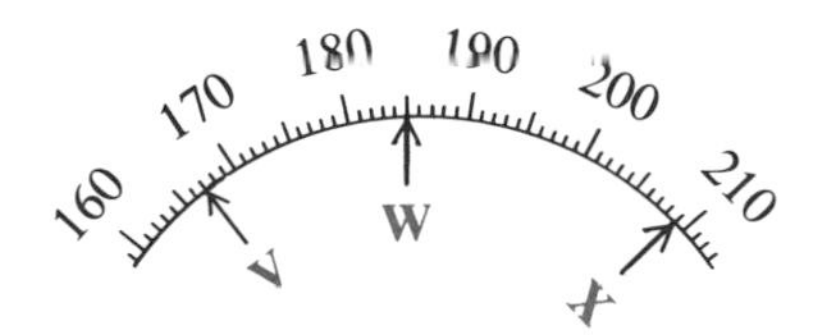

500 510 520 530

Y Z

2

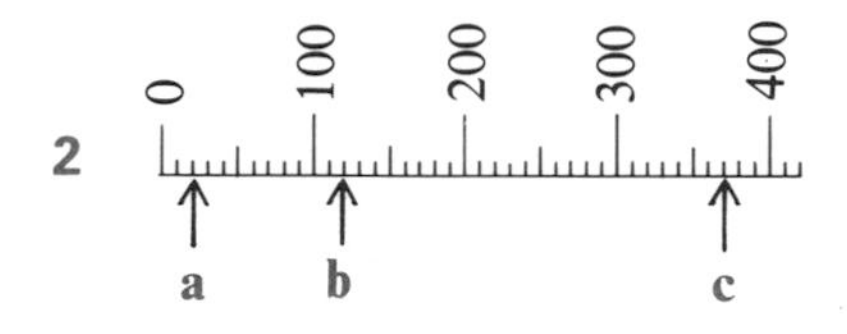

800 900 1000 1100 1200

d e f g

1900 2000 2100 2200 2300 2400 2500

h i j k

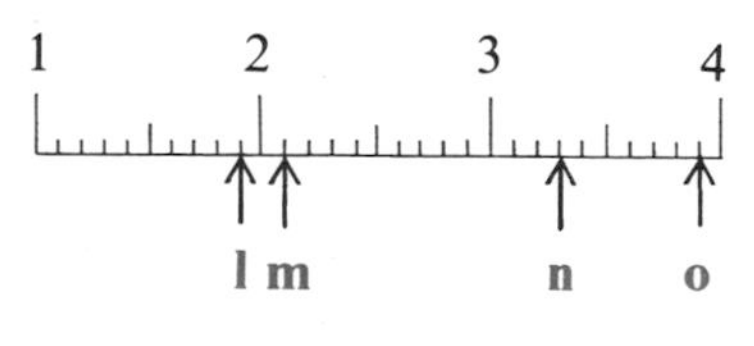

t s p q r

0 0.1 0.2 0.3 0.4

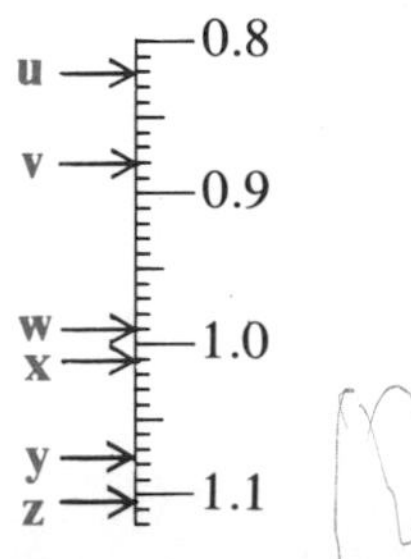

Decimals

Money and Length

Decimals and money

Shopping lists and invoices

Decimals and length

Decimals and money

Part 1 Introduction

How much money is there in each pile? Write your answers in £s.

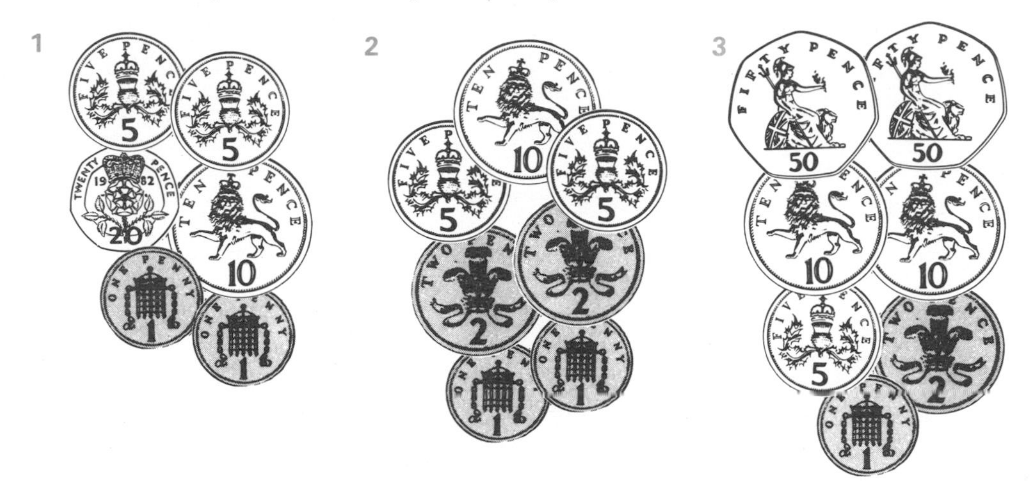

How much do you have altogether with these coins in your pocket?
Give your answers in £s.

4 Eight TENS, two FIVES, three TWOS and two ONES.

5 A FIFTY, four TENS, three TWOS and three ONES.

6 A FIFTY, two TWENTIES, three TENS and two FIVES.

7 Two FIFTIES, three TWENTIES, a FIVE and three TWOS.

8 Three FIFTIES, a TWENTY, three FIVES and a TWO.

9. Eight pupils set off from home with the coins listed here.
How much does each of them have? Give your answers in £s.

a Alan has four FIFTIES, a TWENTY, two TENS and four TWOS.

b Barry has a FIFTY, three TWENTIES, four FIVES, two TWOS and a ONE.

c Carol has five FIFTIES, two TWENTIES, six TWOS and two ONES.

d David has a FIFTY, six TWENTIES and three FIVES.

e Emma has three FIFTIES, four TWENTIES, five FIVES and six ONES.

f Fay has six TWENTIES, eight TENS and five FIVES.

g Gary has three TWENTIES, five FIVES, seven TWOS and six ONES.

h Hannah has a FIFTY, four TWENTIES, three TENS and five FIVES.

Write the values of the **least number** of coins you need to make up these sums of money.

10 £0·60
11 £0·32
12 £0·19
13 £0·25
14 £0·72
15 £0·84
16 £0·12
17 £0·26
18 £0·99

Decimals and money

Part 2 Addition of money

Add these amounts of money.

1	2	3	4	5
£0·32	£0·74	£2·41	£12·72	£ 3·11
0·41	1·31	1·72	8·94	12·42
+1·08	+2·44	+0·53	+20·31	+25·14

6 £1·67 + £3·82 + £9·24

7 £8·07 + £7·42 + £3·87

8 £12·46 + £8·11 + £14·25

9 £28·74 + £31·02 + £7·12 + £0·34

10 £362·40 + £142·71 + £27·36 + £2·47

11 A man pays into his bank four cheques for £62·42, £39·61, £132·50 and £24·76. How much has he paid in altogether?

12 Joanne gets five birthday presents which cost £3·47, £12·14, £1·42, £8·27 and £4·05. How much was spent on these presents altogether?

13 Mr Huggins had to pay four gas bills last year, for £126·42, £86·72, £42·63 and £51·47. How much did he spend on gas during the year?

14 A holiday is advertised at £246·50 per person, but health insurance is £6·45 extra, luggage insurance is £2·85 extra, and airport taxes are £14·75 extra. What is the total cost of the holiday including these extras?

15 Mrs Edwards spends £16·47 at the supermarket, £10·48 at the greengrocers, £4·73 at the butchers and £0·84 at the bakers. How much did she spend in total?

Add the following and answer in £s.

16 £6·42 + £1·37 + 52p

17 £20·37 + £8·64 + 37p

18 £1·24 + 85p + 47p

19 £36·42 + £86·49 + 62p + 53p

20 55p + £1·60 + £20·14 + 37p

21 A housewife buys two tins of soup at 21 pence each, one packet of cereal for 59 pence, two loaves of bread at 42 pence each, a dozen eggs for 75 pence, and a joint of meat for £4·27. How much has she spent altogether?

22 Five children go on a day trip and they take 65p, 52p, £1·32, 87p and £1·05 as pocket-money. How much do they have altogether?

23 Grandmother receives five boxes of chocolates at Christmas. They cost £1·24, 95p, 67p, £1·05 and 72p. What is the total amount spent on Grandma's chocolates?

24 A family go away for the weekend by car and they spend these amounts on petrol: £6·26, £8·58, £10·37 and £4·21. How much is this in total?

25 A gardener visits a local nursery and buys two shrubs costing £1·45 and £2·35 each, thirty daffodil bulbs which cost 24p for ten, an apple tree costing £7·55 and two bags of bone-meal costing 37p each. What amount was he charged?

Decimals and money

Part 3 Subtraction of money

1	2	3	4	5
£6·74	£8·55	£12·76	£43·94	£30·57
−2·51	−4·74	− 7·18	−18·27	−12·85

6	7	8	9	10
£50·74	£40·89	£6·08	£8·07	£4·02
−24·91	−32·94	−4·79	−3·48	−1·57

11	12	13	14	15
£3·34	£6·24	£9·35	£46·25	£74·60
−1·76	−2·38	−4·56	−28·57	−58·72

16	17	18	19	20
£81·64	£24·33	£34·65	£20·32	£40·06
−63·85	−15·89	−29·88	−17·49	−29·49

21 A boy has saved £12·57 and he decides to buy an L.P. record costing £3·86. How much will he have left?

22 Mrs Duncan's bank account has £137·65 in it. If she spends £49·38 on a new coat, how much has she left?

23 A television set is advertised at £148·20. Mr Skelwith buys it with cash and £14·82 is deducted from the price. How much does he pay?

24 A telephone bill consists of the charge for the hire of the equipment plus the cost of the calls made. If the total bill is £27·44 and the hire charge is £8·75, how much was spent on the calls?

25 Mrs Wetherby's son owes her £12·40. Each week he pays part of his debt. In the first month he repays £1·42, £2·36, £0·75 and £1·37. How much does he still owe his mother?

26 A new car costs £4475 and each year it is worth less and less. In the first year it loses £755 of its value; in the second year it loses £325 more; and in the third year £216 more. How much is it worth after three years?

27 Mr Marple earns £425·34 in one month. From this, £65·36 is deducted in tax, £8·76 is taken away for insurance and £18·74 goes to his pension scheme. How much does he have left?

28 The total fuel bills for a house in December are £30·74. The gas bill was £14·54; the electricity bill was £6·27. How much was spent on coal?

Subtraction by 'adding on'

29	30	31	32	33
£5·00	£6·00	£2·00	£1·00	£9·00
−4·94	−5·87	−1·69	−0·77	−7·98

34	35	36	37	38
£7·00	£10·00	£10·00	£4·00	£2·00
−5·89	− 4·94	− 0·84	−2·79	−0·88

39	40
£12·00	£11·00
− 9·71	− 9·59

41 You buy vegetables costing 78p. What change do you get from £1?

42 Mrs Olson went into a shop with a £5 note and left with £3·81. How much did she spend?

Decimals and money

43 A man pays £2·84 for some petrol. What change does he receive from £3?

44 Oxfam had hoped to raise £300 from a flag-day in the village of Cudthorpe. They only received £277·50. By how much were they short of their target?

45 A woman paid for her groceries with a £5 note and she was given £2·84 change. How much did the groceries cost?

46 Mr Murray had a £10 note in his wage packet. On his way home he bought a bottle of wine for £1·77 with this note. How much change did he get?

Part 4 Multiplication of money

1	£2·34 × 2	2	£3·64 × 4	3	£1·23 × 6	4	£3·25 × 4	5	£4·03 × 8
6	£2·15 × 3	7	£3·21 × 5	8	£12·24 × 7	9	£13·44 × 5	10	£25·47 × 6
11	£3·24 × 13	12	£4·32 × 24	13	£6·07 × 34	14	£12·23 × 25	15	£20·36 × 32
16	£41·42 × 16	17	£23·07 × 145	18	£124·21 × 132	19	£207·42 × 227	20	£134·48 × 315

21 If one chair costs £32·45, how much will four chairs cost?

22 The cost for one passenger to cross the English Channel to France is £12·45. What is the charge for six adult passengers?

23 One litre of petrol costs £0·36. How much will it cost to fill an empty tank in a car which holds 45 litres?

24 One square metre of carpet costs £5·74. How much will it cost to carpet a room of 18 m^2?

25 Daffodil bulbs are £0·35 for ten. How much will 140 bulbs cost?

26 On average, Priscilla the hen lays six eggs every week. If eggs sell for £0·66 per dozen, how much will Priscilla have earned in 30 weeks?

27 A man earns £1·56 every hour. How much does he earn in a working week of 36 hours?

28 Curtain material costs £4·55 per metre. What is the cost of a piece which is 24 metres long?

29 Mrs McCann buys three tins of cocoa at £1·15 each and six bars of chocolate at 14 pence each. How much has she spent altogether?

30 Mr Ogden bought five litres of petrol at £0·36 per litre and six litres at £0·38 per litre. How much did he spend on petrol?

31	£4·24 × 12	32	£8·36 × 26	33	£11·52 × 121	34	£25·13 × 235	35	£16·57 × 423

36 If one litre of cooking oil costs £1·26, how much will 12 litres cost?

37 A man earns £1·62 per hour. What is his wage for a 42-hour week?

38 The train fare from Nawood to Rystone is £2·56. If a businessman makes this journey twice a week for sixteen weeks, how much does it cost him?

Decimals and money

Part 5 Division of money

1 Divide £6·84 by 2
2 Divide £9·63 by 3
3 Divide £7·86 by 2
4 Divide £6·08 by 4
5 Divide £17·25 by 5
6 Divide £20·52 by 6
7 Divide £35·28 by 7
8 Divide £20·32 by 4
9 Divide £23·66 by 7
10 Divide £43·65 by 9
11 Divide £7·56 by 6
12 Divide £9·44 by 4
13 Divide £76·64 by 8
14 Divide £75·18 by 6
15 Divide £110·16 by 4
16 Divide £6·42 by 3
17 Divide £16·75 by 5
18 Divide £49·42 by 7
19 Divide £108·84 by 3
20 Divide £172·27 by 7

21 Five boxes of chocolates cost £6·75. How much does one box cost?
22 Three pairs of similar shoes cost £23·52. How much will one pair cost?
23 Mrs Clough buys eight fruit trees for £28·64. If they each cost the same amount, how much was paid for one?
24 A package holiday for six adults cost £574·50. How much is this per person?
25 John delivers newspapers six days of the week and earns £2·70. How much is this per working day?
26 A pack of six tins of fizzy drinks costs £1·08. How much is paid for one tin?
27 Seven litres of petrol cost £2·59. How much does one litre cost?
28 In eight weeks Mrs Ingham spent £14·56 on dog food. How much is this per week? How much is this per day?
29 A music lover spent £23·10 on six equally-priced records. How much were they each?
30 A table and four chairs are advertised for £199·50. The table alone is worth £52·50. How much is one chair sold for?

31 Divide £20·10 by 15
32 Divide £30·42 by 13
33 Divide £73·92 by 21
34 Divide £72·96 by 12
35 Divide £42·56 by 14

36 A man earns £5608·20 in a year. On average, what does he earn per month?
37 Jimmy Hirst went on holiday for two weeks and spent £18·20. How much did he spend on average per day?
38 A man plants a hedge of conifer trees using 35 plants which cost him £92·75. How much was he charged for each tree?
39 On a youth-hostelling weekend, a party of school children share a box of 72 chocolate biscuits costing £5·76. How much is each biscuit?
40 Mrs Clegg earns £5140·20 in a year. What are her average weekly earnings? What are her average daily earnings for a five-day working week?
41 A firm buys petrol in bulk for its cars and pays £129·50 for 350 litres. How much per litre is the firm charged?
42 If it costs £1072·50 to cross the 3250 miles of the Atlantic by Concorde, how much is this per mile?

Decimals and money

Part 6 A mixture of problems

1 During the four weeks of May, Alison received £5·24, £3·62, £2·86 and £4·29 as pocket-money. How much did she receive altogether?

2 In one week Mr and Mrs Bradshaw spend £24·70 on food, £85·60 on their mortgage, £18·16 on petrol and £18·74 on heating. What is the total cost of these items?

3 Deryk went into the sports shop with £15·64 and came out with £8·28 and some sports equipment. How much had he spent?

4 A bank account has £264·85 in it, before £82·46 is drawn out. How much is left?

5 Mr and Mrs Yeates spend £132·60 each month on their mortgage. How much is this in one year?

6 a One car tyre costs £27·45. How much does Mr Wilband spend on four tyres?
b He also buys six litres of oil costing 96p per litre. How much does he spend on oil, and what is his total spending on oil and tyres?

7 A bicycle costing £122·56 on hire purchase is paid for in eight equal instalments. How much is each instalment?

8 Four men travel to work in the same car and share the cost of petrol. During one month the petrol bill came to £30·56, so how much did each man pay?

9 Deborah has £12·42 and she is given another £3·65 so that she can buy a radio. If the radio of her choice costs £14·38, how much will she have left?

10 What will be the total cost if you buy eight bottles of wine costing £2·85 each and six pounds of cheese at £1·24 per pound?

11 Mr Retford had £156·73 in the bank. One day he took out £43·78 but paid in a cheque for £23·94. How much did he then have in the bank?

12 A couple bought a suite of furniture for £449·50 on hire purchase. They paid £150 immediately, £43·50 after one month, and £25·60 after two months. How much did they still owe?

13 The price for a standard car is £3750. A heated rear window is £14·50 extra, a radio is £26·50 extra and radial tyres are £28 extra. Mr Tilson trades in his old car and is given £1355 for it. How much does he have to pay for the new car including all the extras?

14 The train fare from Alston to Barlby is £6·54 for an adult. Children under 3 years go free, and children between 3 and 14 go for half fare. How much does it cost Mr and Mrs Batsby for a return trip with their children Jane (age 2 years) and Stephen (age 9 years)?

15 Mrs Timms spends £2·54 on groceries and £1·24 on meat in the same shop, and pays with a £5 note. She gives the assistant two '10p off' tokens and three '7p off' tokens when she pays. How much change will she get?

16 Mrs Peace takes her two children to visit their grandmother. They have to change buses on the way. It costs the mother 84p on the first bus and 53p on the second one; and the children travel at half fare. What is the cost of making the return trip?

17 I pay a deposit of £41·50 followed by twelve monthly instalments of £12·42 in order to buy a camera. How much did it cost?

18 Miss Rickett's four gas bills for the year cost her £76·42, £52·74, £36·84 and £61·64. If she saves up the same amount each month to pay these bills, how much should she save?

Decimals and money

19 Mr Perman lets cars park on his land, but he charges them 85p for a morning, £1·15 for an afternoon and £1·45 for all the day. If six cars park there in the morning, twelve in the afternoon and eight all day, how much does he collect in charges?

20 The geography department in a school buys 32 copies of a book costing £4·55 per copy, and 64 copies of another costing £1·24 each. How much has the department spent altogether?

21 A train commuter buys a season ticket for £114·50. If he makes 112 journeys using this ticket and each journey would have cost him £1·34, how much has he saved by buying the season ticket?

22 The cash price of a hi-fi set is £247. If I buy it on hire purchase with a deposit of £52 followed by six monthly instalments of £37·75, how much extra have I to pay?

23 In one month of 30 days, my dog eats $1\frac{1}{2}$ tins of food at 25p per tin each day and also two bags of biscuits at £1·27 each. How much did it cost per day to feed my dog?

24 I had £247·52 in my bank account. If I drew out £64·75 but paid in a cheque for £124·62, how much did I then have in the account?

25 On her birthday Diana is given £5 by her Aunt Mary, £3 by her Uncle George, £8 by her mother and £2·50 by her brother. If she buys new clothes for £21·45, how much had she to provide herself?

26 Twelve people go on a camping holiday and pay £65·40 each. If the total cost of the holiday is £621·60, how much refund should they each have?

27 Fifteen school children go on a trip costing £14·50 each. If the total cost proves to be £184·95, how much should they each be refunded?

28 Mr Godber is paid 14p per mile when he uses his own car on business. In one month he travels 249 miles, but loses one third of his payment in tax. How much does he receive after the tax is deducted?

29 Mrs Kine is paid 15 pence per mile when her car is used for the firm's business. But she loses $\frac{1}{4}$ of this in tax. If she travels 136 miles, how much will she receive after tax is deducted?

30 Miss Chay blends her own tea and sells it to her friends without making any profit. She takes 2 kg of one type of tea costing £2·59 per kg and mixes it with 6 kg costing £2·15 per kg. How much does she charge for each $\frac{1}{2}$ kg packet of the blend?

Shopping lists and invoices

Find the total amounts of money required to purchase the following lists of goods. Give all your answers in £s.

1

4 pkts of butter at 42p/pkt =
3 kg of sugar at 38p/kg =
2 loaves of bread at 41p/loaf =
8 kg of potatoes at 18p/kg =
6 kg of carrots at 31p/kg =

TOTAL £

2

5 tins of tomatoes at 21p/tin =
3 tins of peas at 15p/tin =
8 tins of condensed milk at 11p/tin =
4 kg of flour at 34p/kg =
2 bottles of oil at 78p/bottle =

TOTAL £

3

3 litres of oil at 92p/litre =
20 litres of petrol at £0·35/litre =
2 tyres at £16·35 per tyre =
3 doz screws at 32p/doz =
2 headlight bulbs at £1·62 each =

TOTAL £

4

5 litres of paint at £1·86/litre =
4 litres of undercoat at £1·67/litre =
3 paintbrushes at 48p each =
12 rolls of wallpaper at £2·35 each =
2 pkts of paste at 52p each =

TOTAL £

5

4 pkts of envelopes at 45p each =
3 reams of paper at £1·37 each =
2 writing-pads at 52p each =
15 pencils at 12p each =
16 ball pens at 18p each =
2 boxes of pins at 23p/box =

TOTAL £

6

2 spades at £6·48 each =
3 shrubs at £1·72 each =
4 trees at £3·85 each =
600 g of grass seed at 26p/100g =
8 pkts of flower seeds at 28p/pkt =
$4\frac{1}{2}$ doz bulbs at 68p/doz =

TOTAL £

7

2 boxes milk chocolates at £4·52 each =
5 boxes plain chocolates at £5·24 each =
1 doz Krunshy bars at 15p each =
$\frac{1}{2}$ doz boxes of Smartlets at 68p/box =
16 bars of nougat at 14p/bar =
$3\frac{1}{2}$ doz Kreem Eggs at £1·68/doz =

TOTAL £

8

4 men's shirts at £8·65 each =
2 pairs of shoes at £12·84/pair =
$2\frac{1}{2}$ doz handkerchieves at £5·86/doz =
14 pairs nylon socks at 64p/pair =
8 pairs wool socks at £1·24/pair =
3 pairs gents trousers at £14·42/pair =

TOTAL £

Shopping lists and invoices

9

6 passengers paying 15p each =
4 passengers paying 18p each =
10 passengers paying 17p each =
22 passengers paying 24p each =
8 passengers paying 20p each =
3 passengers paying 22p each =

TOTAL £

10

65 pkts exercise books at £1·34/pkt =
82 junior textbooks at £1·82 each =
125 middle-school books at £2·05 each =
12 sixth-form books at £3·14 each =
6 boxes of chalk at £1·76/box =
86 rulers at 11p each =
8 doz pencils at 52p/doz =

TOTAL £

11

4 tins of soup at 17p each =
2 bars soap at 18p each =
$\frac{1}{2}$ kg butter at 98p/kg =
6 boxes of matches at 4p each =
2 bottles of milk at 21p each =
$\frac{1}{2}$ doz eggs at 62p/doz =

TOTAL £

12

6 balls of wool at 25p/ball =
4 pairs of needles at 52p/pair =
2 patterns at £1·27 each =
4 pkts of buttons at 23p/pkt =
2 zips at 72p each =
3 pkts needles at 42p/pkt =

TOTAL £

13

$2\frac{1}{2}$ m^2 of glass at £5·80/m^2 =
4 pkts of tacks at 22p/pkt =
$6\frac{1}{2}$ m of beading at £0·46/m =
2 hack-saw blades at 18p each =
$3\frac{1}{2}$ doz screws at 28p/doz =
5 sheets sandpaper at 12p/sheet =

TOTAL £

14

2 kg of flour at 36p/kg =
$\frac{1}{2}$ kg of sultanas at £1·06/kg =
$\frac{1}{2}$ kg of sugar at 42p/kg =
2 pkts glacé cherries at 15p/pkt =
$\frac{1}{2}$ litre of milk at 44p/litre =
$1\frac{1}{2}$ doz eggs at 68p/doz =

TOTAL £

15

$\frac{1}{2}$ kg of liver at £1·66/kg =
2 kg of bacon at £1·78/kg =
$1\frac{1}{2}$ kg of mince at £1·86/kg =
3 kg of beef at £2·82/kg =
$4\frac{1}{2}$ doz eggs at 72p/doz =
$\frac{1}{2}$ doz meat pies at 36p/each =

TOTAL £

16

14 newspapers at 15p each =
25 comics at 13p each =
9 magazines at 22p each =
12 paperbacks at 85p each =
4 hardbacks at £2·63 each =
5 booklets at 14p each =

TOTAL £

Decimals and length

1 Use a ruler to measure in **centimetres** the distance between the following pairs of points on these lines.

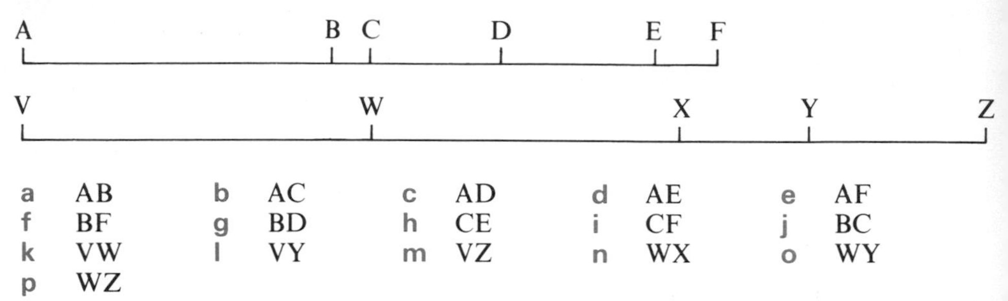

a AB   b AC   c AD   d AE   e AF
f BF   g BD   h CE   i CF   j BC
k VW   l VY   m VZ   n WX   o WY
p WZ

Measure in **centimetres** the lengths of these jagged lines.

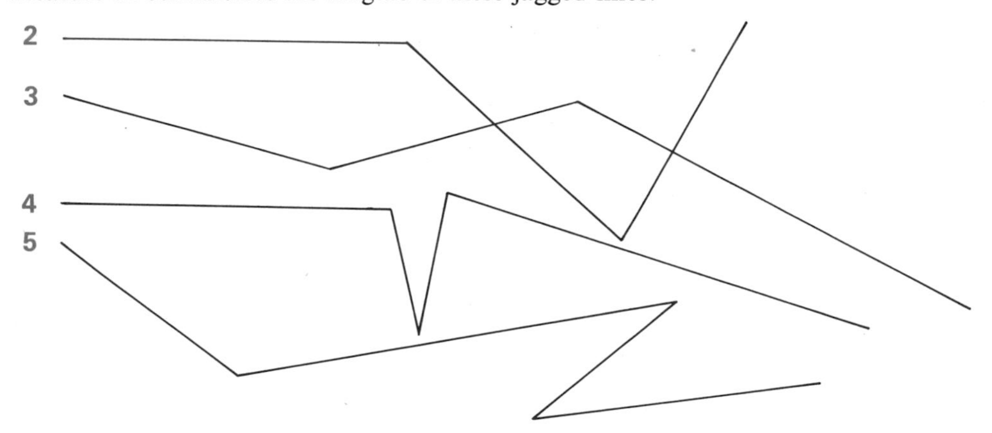

The distance all the way round any shape is called its **perimeter**.

Measure the lengths of the sides of each shape and add them together to find the **perimeters**.

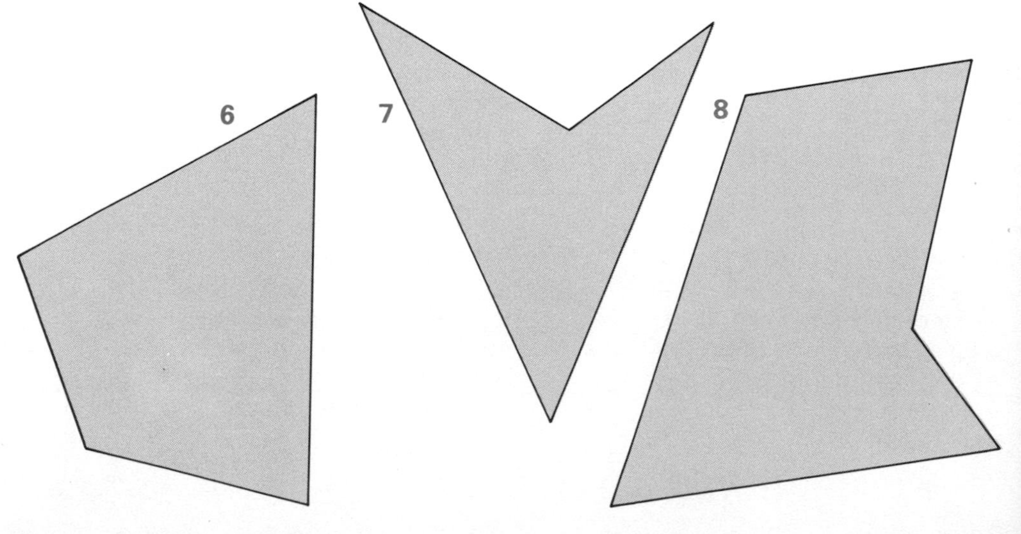

Decimals and length

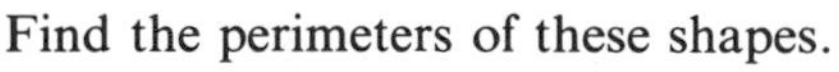

Find the perimeters of these shapes.

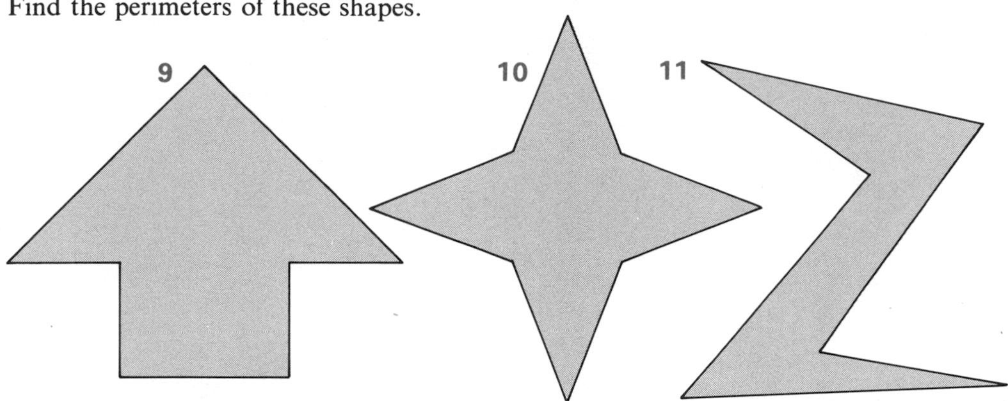

Measure in centimetres the lengths of the following pairs of lines.
How much longer is line A than line B for each pair?

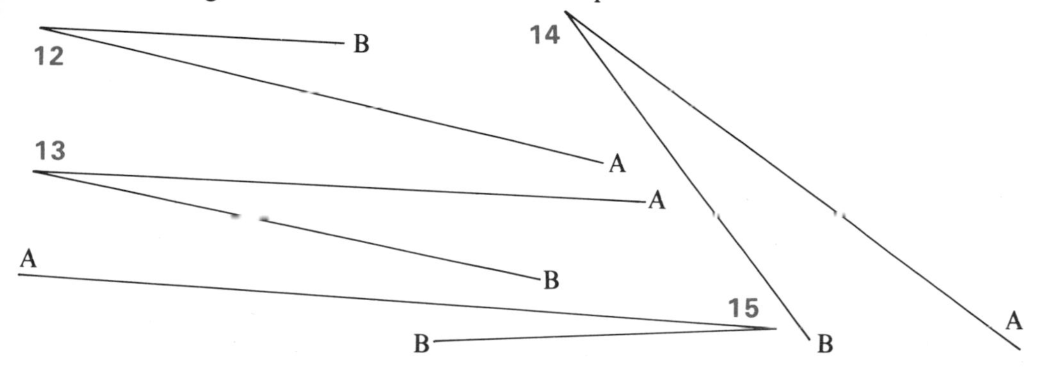

16 Write, in *three* different ways, the distances between the following pairs of arrows.

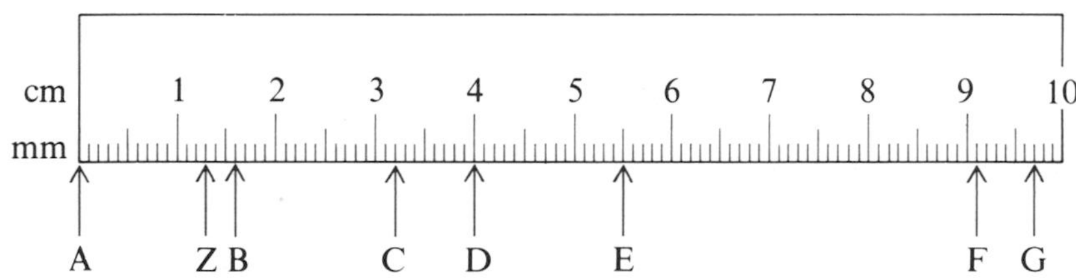

Example AZ = 13 mm = 1.3 cm = $1\frac{3}{10}$ cm

a	AB	**b**	AC	**c**	AE	**d**	AF	**e**	DE	**f**	DF
g	DG	**h**	BD	**i**	CD	**j**	FG	**k**	BC	**l**	EF

17 Use a ruler to measure in **centimetres** and also in **millimetres** the lengths of these lines. Write each length in *three* different ways.

a ————————————————————

b ————————————————————————————————————

c ——————————————————————————————

d ———

e ——

Decimals and length

The following three maps are each drawn to a different scale.
Use a ruler to find the shortest distances between the given places.

18 A village
Scale: 1 cm = 1 km

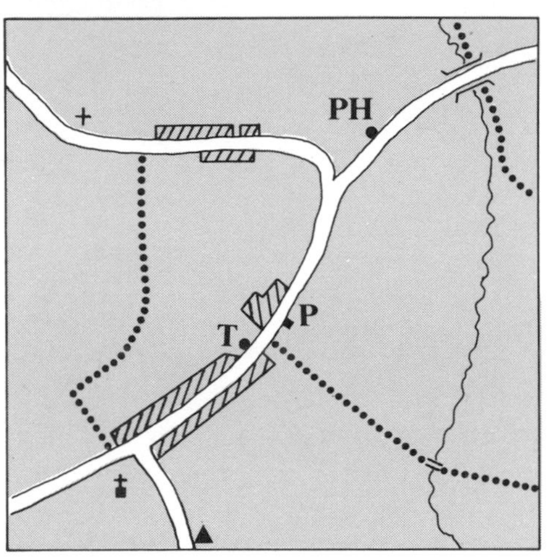

a YOUTH HOSTEL to TELEPHONE
b CHURCH to PUBLIC HOUSE
c POST OFFICE to CHAPEL
d ROAD BRIDGE to FOOT-BRIDGE
e TELEPHONE to PUBLIC HOUSE
f CHAPEL to FOOT-BRIDGE
g POST OFFICE to PUBLIC HOUSE

19 Towns and villages
Scale: 1 mm = 1 km

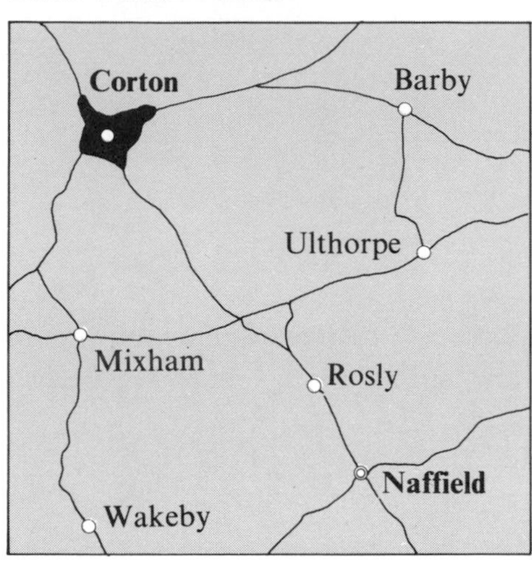

a Mixham to Ulthorpe
b Barby to Naffield
c Barby to Wakeby
d Mixham to Rosly
e the centre of Corton to Naffield
f the centre of Corton to Wakeby
g Ulthorpe to Rosly

20 Southern Italy
Scale: 1 mm = 10 km

a Rome to Reggio
b Rome to Taranto
c Naples to Rome
d Taranto to Palermo
e Palermo to Rome
f Reggio to Naples
g Naples to Taranto

Decimals and length

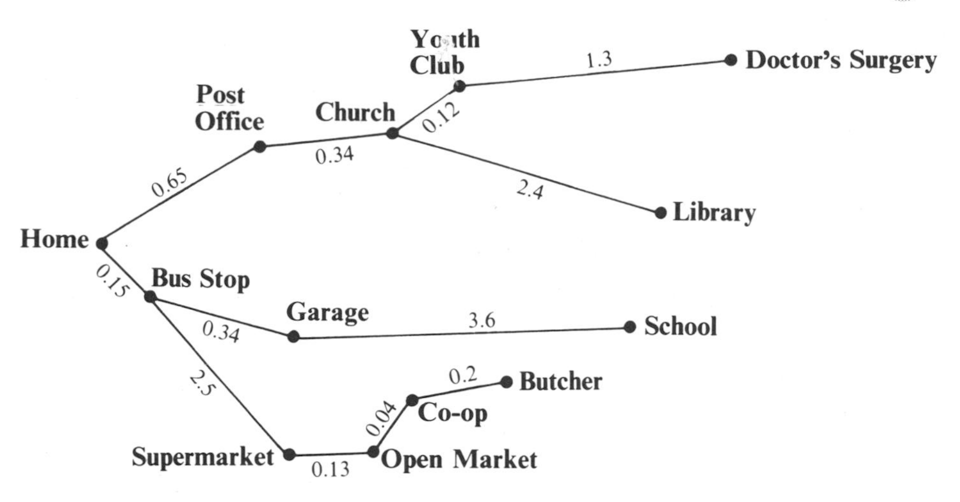

21 The above map shows distances in **kilometres** from your home to various local places. It is not drawn to scale.

By adding distances (do *not* use a ruler), find how many kilometres are between your **home** and the following places.

a Church
b Youth Club
c Doctor's Surgery
d Library
e Supermarket
f Co-op
g School
h Butcher

By subtracting distances, find how much further from your **home** is:

i the Post Office than the Bus Stop,
j the Church than the Garage,
k the Supermarket than the Youth Club
l the School than the Library.

Also find m how much further from the Bus Stop is the Supermarket than the Garage;
n how much further from the Church is the Library than the Surgery.

Addition and subtraction

22 Patrick went with his family on a four-day holiday by car. He noted the number of miles travelled each day: 267.5, 37.3, 13.8 and 257.9. Find the total number of miles they travelled.

23 Find the height of a pile of 5 books if they are 5.6cm, 4.3cm, 3cm, 0.8cm and 2cm thick.

24 Three pieces of electrical wire are joined together. If the separate pieces are 4m, 2.5m and 1.25m long, find their total length.

25 During three successive weeks of no rain, the level in a reservoir fell by 0.85 m, 1.23 m and exactly 2 m. Find the total fall in the level.

26 Jane cycles the 15.4 km to her grandmother's house. If 4.8 km of the journey are on main roads, what distance is on minor roads?

27 A room is 6.3 m long and a carpet covers 4.9 m of this length. What length of the room is not covered by the carpet?

Decimals and length

28 Dave Kershaw is a long-distance lorry driver. He set off from London with the odometer showing 67 175.8 km. Five days later, in Athens, the meter showed a reading of 70 384.3 km. How far did he drive from London to Athens?

29 Find the length XB if
- **a** AB = 16.3 cm and AX = 4.8 cm
- **b** AB = 16.3 cm and AX = 7.7 cm.

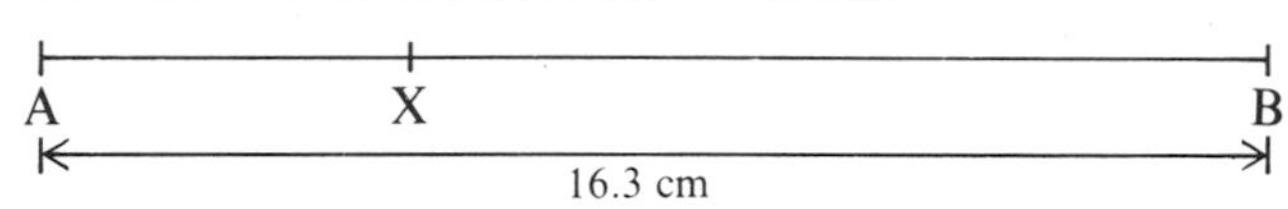

30 The string in a new ball is 25 metres long. If 13.7 m is cut off, how much is left?

31 A building has a flag-pole on its roof. If the total height of both is 32 metres but the building alone is 28.6 m high, what is the height of the flag-pole?

Multiplication and division

32 Butter is delivered in rows of packets with 6 packets per row. If each packet is 7.8 cm across, how long is a row?

33 Goods are stored in a warehouse in piles of boxes with 12 boxes in a pile. If all the boxes are the same size and each is 0.85 m high, what is the height of a pile?

34 A square has sides 23.8 cm long. How long is the perimeter of the square?

35 One lap in a cross-country cycle race is 5.3 km long. If the race has 15 laps, how far do the cyclists have to travel to finish?

36 If the teeth on a comb are 0.15 cm apart and there are 76 teeth on it, find the length of the comb.

37 Mr Batley drives 12.7 miles to and from work every day. How far does he travel
- **a** in a working week of five days,
- **b** in a year of 46 working weeks?

38 On one litre of petrol, a car will travel 15.6 km. How far will it travel on
a 12 litres **b** 13.5 litres?

39 A garden path is made from eight identical paving stones placed in line. If the path is 6.56 m long, find the length of each paving stone.

40 A piece of string 6.25 m long is cut into 5 equal lengths. How long is each bit?

41 A motor scooter travelled 263.2 km on 7 litres of petrol. How far would it go on one litre?

42 Travelling at a steady speed on a motorway, a lorry covered 306 km in exactly 4 hours. How far did it average in one hour?

43 A baby's book has eight cardboard pages. If the book is 3.4 cm thick, find the thickness of one page.

44 A paperback book is made from 80 sheets of paper having a total thickness of 6.8 mm. What is the thickness of just one sheet?

45 Light travels a distance of 186 000 miles every second. How far does it travel in one year? (This distance is called a *light-year*.)

Decimals

General Exercises

Addition of decimals
Subtraction of decimals
Multiplication of decimals
Division of decimals
Decimal problems
Multiplying and dividing by 10, 100 etc.
Change of units
Revision
Approximations

Addition of decimals

Work the following.

1	2	3	4	5
123.1	67.14	6.781	45.9	801.4
245.0	12.3	11.8	1.201	32.01
+ 61.1	+ 1.45	+ 2.03	+32.74	+1102.5

6 6.245 + 2.314 + 1.029

7 12.55 + 6.2 + 1.07

8 65.99 + 1.3 + 44.0

9 16.23 + 4.61 + 23.084

10 145.6 + 3.02 + 1.542

11 85.6 + 145.23 + 1.008 + 34.1

12 6.425 + 0.75 + 8.0

13 A salesman travels 56.7 miles on Monday, 89.6 miles on Tuesday, 345.8 miles on Wednesday and 106.6 miles on Thursday. How far does he travel altogether?

14 A woman pays four cheques into her account at the bank for £6·72, £12·60, £63·48 and £234·75. How much does she pay in altogether?

15 John has saved £15. He has £0·46 in his pocket and will be paid £1·50 tomorrow. How much will he then have?

16 A length of cable is cut into three pieces which are 6.4 metres, 1.27 metres and 0.33 metres long. How long was the cable before it was cut?

17 A housewife buys 1.25 kg of sugar, 0.75 kg of tea, 0.275 kg of cheese and 2.5 kg of flour. What is the total mass of her purchases?

18 A girl wraps a parcel and uses four pieces of sellotape. They are 2 metres, 1.2 metres, 0.65 metres and 0.08 metres long. How much tape did she use altogether?

19 Mrs Black took her dog for three walks on Saturday. The first walk was 1.2 km long, the afternoon walk was 6.75 km long, and the evening walk 3 km long. How far did she walk that day?

20 A rectangle is 8.25 cm long and 3.4 cm wide. Find its perimeter (the distance round the edge of the rectangle).

Work the following.

21 0.035 + 0.064 + 0.43

22 0.875 + 0.032 + 0.006

23 0.0057 + 0.64 + 0.078

24 0.932 + 0.0045 + 0.24

25 0.0006 + 0.084 + 0.07 + 0.408

26 In a science lesson, the teacher puts 0.065 kg of powder and 0.15 kg of acid into a beaker with a mass of 0.082 kg. What is the total mass of the beaker and its contents?

27 A young boy cycles to his grandmother's house and stops three times on the way. Between stops he travels 0.085 km, 0.16 km, 2.5 km and 6 km. What is the total distance he cycles?

28 Mr Brown has a heater which burns paraffin. Over five days, it used 8 litres, 2.5 litres, 0.75 litres, 3 litres and 0.5 litres. How much paraffin is this altogether?

29 Madame Legrange bought 3 kg of potatoes, 0.4 kg of peppers, 2.5 kg of carrots and 0.175 kg of mushrooms. What is the total mass of these purchases?

30 In a kitchen there are three containers of washing-up liquid holding 4 litres, 0.65 litres and 2.7 litres. What is the total volume of washing-up liquid?

Addition of decimals

31 A building 34.6 m high has a flag-pole 4.25 m high on its roof. To the top of the flag-pole is fixed a lightning-conductor 0.45 m tall. How far above ground level is the top of the lightning-conductor?

32 Find the perimeter of triangle ABC (i.e. the distance round its edge).

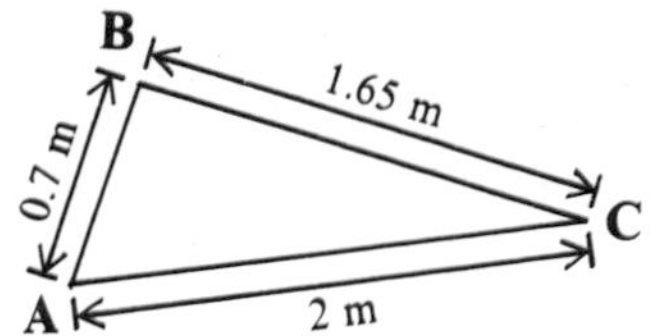

33

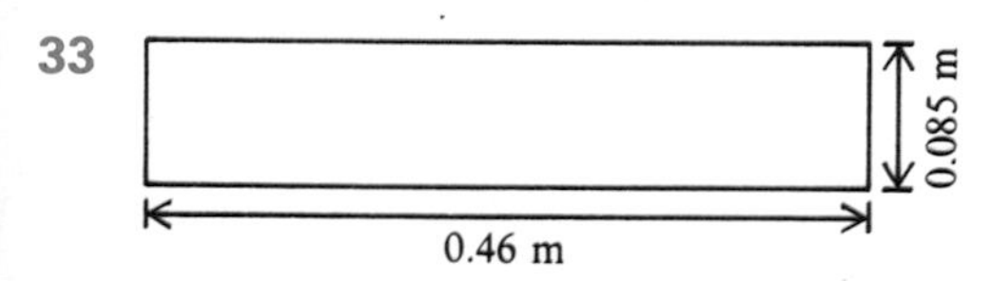

Find the perimeter of this rectangle.

34 A rectangle is 0.27 m long and 0.124 m wide. Find its perimeter.

35 This pentagon has two pairs of sides equal in length. If x = 0.65 metres, y = 2.3 metres and z = 2 metres, what is the perimeter of the pentagon?

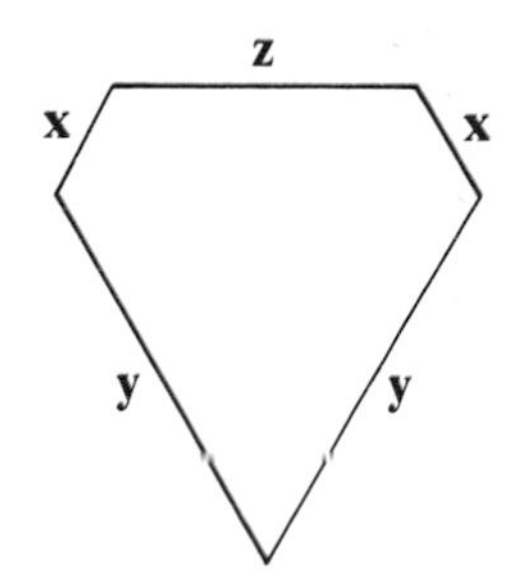

Subtraction of decimals

Work the following.

1	42.4 −31.2	**2**	57.4 −25.6	**3**	6.57 −1.93	**4**	450.6 −316.5	**5**	0.768 −0.279
6	64.2 −38.3	**7**	8.674 −1.925	**8**	371.8 − 92.5	**9**	0.067 −0.019	**10**	87.3 − 9.1

11 If 3.6 metres are cut off a piece of tape 5.4 metres long, what length is left?

12 Mrs Walton had £8·42 in her purse. When she got home from the shops, she had £1·65 left. How much had she spent?

13 Mr Spalding travels 18.5 km to work. If he goes 13.8 km by bus, how much further has he got to go?

14 A man of mass 106.5 kg started to diet. If he loses 34.8 kg, what is his mass now?

15 A tin of cooking oil holds 12.4 litres. If 9.6 litres are poured out, how much is left?

Work the following.

16	46.3 −23.15	**17**	85.4 −37.26	**18**	4.67 −1.828	**19**	0.62 −0.317	**20**	0.82 −0.655
21	4.7 −2.86	**22**	9.2 −4.58	**23**	46 −23.4	**24**	67 −14.9	**25**	158 − 25.2

26 A car starts a journey with a full tank of 37 litres. At the end of the journey there are only 12.9 litres left. How much petrol has been used?

27 A 3.4-metre strip is cut off a 12-metre roll of sellotape. How much is left?

28 Mr Jones has £376 in his wallet. If he spends £184·50, how much has he left?

29 A bag of animal food has a mass of 88 kg. If 47.5 kg are taken out, what mass of food is left in the bag?

30 The tank which holds oil for the central heating in a house takes 545 litres when full. If 127.6 litres are left in the tank, how much of the full tank has been used?

Work the following.

31	20.7 −14.9	**32**	40.2 −21.7	**33**	6.03 −4.27	**34**	0.605 −0.176	**35**	0.503 −0.116
36	40.0 −15.6	**37**	60.0 −37.4	**38**	3 −1.65	**39**	12 − 8.43	**40**	17 − 3.82

41 From a bottle holding 10 grams of a chemical, 1.74 grams are taken out. What mass of chemical is left?

42 A roll of cloth is 10 metres long. A customer buys a length of 4.65 metres. What length is left?

43 A meal in a restaurant costs £8·57 for two people. If the bill is paid with a £20 note, what change will there be?

44 A bag of sugar should have a mass of 2 kg but one particular bag is 0.14 kg too light. What is the mass of this bag?

45 A 100-metre race is run on a track which is 0.35 metres too short. What is the actual length of this track?

Multiplication of decimals

Part 1 Whole numbers

1	123×3	2	142×4	3	267×5	4	328×4	5	1356×3
6	123×43	7	142×34	8	267×35	9	328×14	10	1356×43
11	135×242	12	324×153	13	1305×223	14	2044×342	15	427×203

16 One kilogram of apples costs 67 pence. How much would 13 kilograms cost? How much change would you receive from a £10 note?

17 The flight from London to New York takes 7 hours, and the aircraft averages a speed of 457 miles per hour. Find the distance from London to New York.

18 A girl earns 85 pence in one hour. How much does she earn in 36 hours?

19 One metre of dress material costs 76 pence. How much do 14 metres cost?

20 Every minute 528 litres of water flow out of an open tap. If it takes 45 minutes to fill a water tank, how many litres does the tank hold?

21 A steel spring stretches 12 cm for every kilogram it is holding. If 24 kg is put on the spring, how far will it stretch?

22 A packet of breakfast cereal has a mass of 645 grams. If a supermarket shelf holds 48 such packets, what mass of cereal is on it?

23 The area of one sheet of paper in a book is 580 cm^2. If the book has 630 pages **a** how many sheets are needed **b** what is their total area?

24 An aeroplane takes off with 124 people aboard. If each person has the maximum allowable mass of luggage of 20 kg, what is the total mass of luggage in the aeroplane?

25 Every minute, 135 m^3 of water flow through a pipe from a reservoir. What volume of water flows through **a** in one hour **b** in one day?

26 A car driver averages 47 miles per hour on a journey from London to Athens. If his driving time is 43 hours, how far is it between these cities? If he takes 10 minutes rest for every hour of driving, what would his total time be?

Part 2 Whole numbers and decimals

1	12.3×3	2	4.24×2	3	56.3×5	4	3.64×16	5	25.4×23
6	4.03×54	7	0.136×37	8	0.274×55	9	16.5×232	10	5.04×437

11 A small car averages 10.4 miles to one litre of petrol. How far will it travel on 48 litres?

12 A family car averages 7.2 miles per litre. How far will it go on 54 litres?

13 Curtain material costs £1·55 per metre. How much will 15 metres cost?

14 A yacht travels at a steady 12.5 nautical miles per hour. How far will it travel in 2 days?

Multiplication of decimals

15 £1 can be changed for 2.04 U.S. dollars. How many dollars will you get for £57?

16 A builder uses 4.8 m^2 of formica in every kitchen he builds. If he builds an estate with 86 houses, how much formica will he need?

17 On average, a school uses 42.5 units of gas every hour. How much gas is used during the day from 9 a.m. to 4 p.m. (including the lunch hour)?

18 A gardener sprinkles 12 grams of grass seed on one square metre of garden. How much seed will he need for 82.5 m^2?

19 In May, the Indian town of Skwelshipur averages 3.2 cm of rain every day. What is the total rainfall expected during May?

20 If a crate of bananas sells for £5·74, how much will be charged for 14 such crates?

21 £1 can be changed for 4.71 German marks. How many marks will you get for
 a £65 b £216?

22 Mr Anderson says he averaged 37.5 miles every hour when he went on a touring holiday by car. If he drove for 5 hours each day, how far do you expect he travelled in a one day b a week?

Part 3 Decimals

1 1.23×4.3 2 2.45×3.1 3 47.6×2.3 4 204×5.2 5 3.18×2.4

6 2.45×12.3 7 52.3×4.34 8 6.35×26.7 9 2.44×1.05 10 0.352×12.6

11 3.42×1.77 12 67.3×2.04 13 12.06×34.5

14 0.67×1.52 15 0.523×24.3 16 32.4×0.52

17 A hiker walks at 8.5 km per hour. How far does he walk in 6.5 hours?

18 A rectangle is 62.4 cm wide and 13.6 cm long. Find its area.

19 A boy earns £1·26 each hour. How much does he earn in 4.5 hours?

20 A ball rebounds 0.72 of the height from which it is dropped. If the ball is dropped from 2.65 metres, find the height of its rebound.

21 A man who frames pictures charges £3·75 for every metre length of frame he uses. If a picture needs 1.60 m of frame, what does he charge?

22 A downhill skier averages 12.5 metres every second. If he takes 142.8 seconds during a race, how long is the course?

23 A cricket pitch is protected by a rectangular fence 38.5 m long and 17.6 m across. What area is enclosed by the fence?

24 If £1 can be changed for 1.45 roubles, how many roubles will you get for
 a £12·60 b £0·80?

25 If 1 cm^3 of aluminium has a mass of 2.7 grams, what is the mass of a piece of this metal having a volume of 0.64 cm^3?

26 Between Eckby and Frampton there is 0.35 km of white lines down the middle of every 1 km of road. If it is 12.4 km from Eckby to Frampton, what length of white lines is there between these villages?

Division of decimals

Part 1 Dividing by a whole number

1 Divide 6.72 by 2
2 Divide 49.6 by 2
3 Divide 7.66 by 2
4 Divide 10.62 by 3
5 Divide 4.62 by 3
6 Divide 84.5 by 5
7 Divide 61.3 by 5
8 Divide 21.4 by 5
9 Divide 7.5 by 6
10 Divide 6.34 by 4

11 $27.3 \div 5$
12 $0.544 \div 4$
13 $0.942 \div 6$
14 $0.812 \div 2$
15 $6.18 \div 3$

16 $\frac{4.25}{5}$
17 $\frac{3.708}{12}$
18 $\frac{0.74}{4}$
19 $\frac{0.98}{8}$
20 $\frac{4.34}{8}$

Change these fractions to decimals.

21 $\frac{3}{8}$ 22 $\frac{5}{8}$ 23 $\frac{7}{8}$ 24 $\frac{3}{4}$ 25 $\frac{1}{4}$ 26 $\frac{13}{20}$
27 $\frac{11}{20}$ 28 $\frac{18}{25}$ 29 $\frac{5}{16}$ 30 $\frac{7}{16}$

31 Six records at the same price cost a total of £27·36. What is the cost of each of them?
32 Eight refills for a camping gas stove cost £5·92. Find the cost of one refill.
33 A van is loaded with twelve similar crates of fruit. If the total mass of the load in the van is 160.8 kg, what is the mass of one of these crates?
34 16 tiles are stuck onto 1800 cm^2 of a kitchen wall. What area will just one of these tiles cover?
35 14 tins of paint, each containing the same amount, are emptied into one large container. If there are 17.5 litres altogether, how much paint did each tin hold?

Part 2 Recurring decimals

Work out each of the following until your answers start to recur.

1 Divide 14.5 by 3
2 Divide 26.2 by 3
3 Divide 37.1 by 3
4 Divide 13.4 by 6
5 Divide 67.4 by 6
6 Divide 265.0 by 6
7 Divide 123.4 by 7
8 Divide 14.5 by 9
9 Divide 8.48 by 9
10 Divide 7.69 by 9

11 $36.5 \div 11$
12 $6.87 \div 11$
13 $13.0 \div 12$
14 $2.84 \div 12$
15 $32.5 \div 15$

16 $\frac{1.93}{15}$
17 $\frac{2.0}{7}$
18 $\frac{15.1}{7}$
19 $\frac{0.6}{7}$
20 $\frac{0.1}{7}$

Change these fractions to recurring decimals.

21 $\frac{1}{3}$ 22 $\frac{2}{3}$ 23 $\frac{1}{6}$ 24 $\frac{5}{6}$ 25 $\frac{1}{9}$ 26 $\frac{2}{9}$
27 $\frac{7}{9}$ 28 $\frac{2}{11}$ 29 $\frac{10}{11}$ 30 $\frac{1}{11}$ 31 $\frac{1}{7}$ 32 $\frac{2}{7}$

Division of decimals

Part 3 Dividing by a decimal

Example Divide 4.36 by 0.2 $\frac{4.36}{0.2} = \frac{43.6}{2} = 2\overline{)43.6}$ 21.8 Ans

1 Divide 2.34 by 0.2
2 Divide 4.74 by 0.2
3 Divide 8.82 by 0.3
4 Divide 6.25 by 0.5
5 Divide 6.25 by 0.05
6 Divide 6.25 by 0.005
7 Divide 0.741 by 0.03
8 Divide 0.9632 by 0.004
9 Divide 1.694 by 0.07
10 Divide 3.84 by 1.2

Work the following.

11 $2.88 \div 1.2$
12 $0.6744 \div 0.12$
13 $0.3888 \div 0.012$
14 $4.52 \div 0.08$
15 $7.12 \div 0.5$
16 $0.0318 \div 0.15$
17 $0.3894 \div 1.1$
18 $0.01635 \div 0.025$
19 $0.5334 \div 0.21$
20 $5.84 \div 0.016$

21 a If £2 can be changed for $4.48, how many dollars will you get for £1?
b If £2·40 can be changed for $5.16, how many dollars will you get for £1?

22 a If £2 can be changed for 2.86 roubles, how many roubles will you get for £1?
b If £2·20 can be changed for 3.19 roubles, how many roubles will you get for £1?

23 a If 3 cm^3 of gold have a mass of 57.9 grams, what is the mass of 1 cm^3?
b If 1.5 cm^3 of lead have a mass of 17.1 grams, what is the mass of 1 cm^3?

24 If 2.4 cm^3 of platinum have a mass of 51.6 grams, what is the mass of 1 cm^3 of platinum?

25 If 1.3 cm^3 of diamonds have a mass of 4.589 grams, what is the mass of 1 cm^3 of diamonds?

26 a 5 m^2 of carpet cost £36·80. What would 1 m^2 of this carpet cost?
b 1.5 m^2 of a different carpet cost £6·48. What is the price of 1 m^2?

27 1.2 km of a country lane needs 49.8 tonnes of tarmac to resurface it. How many tonnes will be needed to do 1 km?

28 A farmer uses 3.55 m^3 of insect spray on 2.5 hectares of land. How much spray does he need for 1 hectare?

29 Workmen sow the edge of a road with grass seed. If they scatter 10.6 kg of seed on 0.2 hectares, how much seed would they need for 1 hectare?

30 Harry Gubbins cycles 17.2 km in 0.8 hours. At this rate, how far would he cycle in 1 hour?

31 It takes Mr Wormald 0.4 hours to drive 15.4 km in the city. At this speed, how far would he go in 1 hour?

32 An aeroplane takes 1.2 hours to fly the 750 km between two cities at a steady speed. How far would it go in one hour?

33 A man uses 2.5 litres of petrol to travel 27 miles. How many miles does he travel per litre?

Division of decimals

34 a A pile of paper 2 cm high contains 280 sheets. How many sheets would be in a pile 1 cm high?

b A pile of different paper is 0.45 cm high and contains 54 sheets. How many sheets will be needed for a pile 1 cm high?

35 a A biologist counts 84 cells in an area of 2 cm^2. How many cells would he expect to find in 1 cm^2?

b He then counts under the microscope 26 cells in an area of 0.08 cm^2. How many of these cells would he expect to find in 1 cm^2?

36 If 0.07 hectares of land was bought for £332·64, how much would one hectare cost?

37 If 0.09 cm^3 of silver has a mass of 0.945 grams, what is the mass of 1 cm^3 of silver?

38 If 0.035 cm^3 of bamboo has a mass of 0.0147 grams, what is the mass of 1 cm^3 of bamboo?

39 How many pieces of string 0.15 metres long can be cut from a ball containing 19.8 metres?

40 How many blocks of butter of mass 0.24 kg can be cut from a large tub of mass 56.4 kg?

41 How many bottles holding 0.7 litres of wine can be filled from a vat containing 871.5 litres?

42 How many lengths of cloth 1.4 metres long can be cut from a roll 32.2 metres long?

43 A field of 3.9 hectares is divided into plots having areas of 0.15 hectares. How many plots will there be?

44 How many packets of tea of mass 0.25 kg can be made from a crate of tea of mass 24.5 kg?

45 A bottle of nail varnish holds 25.4 cm^3. If, on average, a woman uses 0.002 cm^3 of nail varnish on each nail, how many nails can be painted using this bottle?

Decimal problems

Addition and multiplication

Part 1

1 Adrian has £2·48 and his sister Julie has £3·85. How much do they have altogether?

2 In the first week of March John saved £6·72, in the second week he saved £2·13, in the third week £4·35, and in the last week £12·08. How much did he save altogether?

3 Mrs Nelson made four purchases costing £6·42, £3·91, £13·25 and £21·52. How much did she spend altogether?

4 On a five-day cycling holiday, a boy travelled 38.7 miles, 69.4 miles, 107.8 miles, 9.6 miles and 41.7 miles. What was the total distance he cycled?

5 Angela travels 5.3 miles to school in the morning. How far does she travel
a each day b each week of five days?

6 If one litre of petrol costs £0·37, how much do
a 2 litres b 13 litres cost?

7 A car leaves Carlisle and travels south on the M6 averaging 62.5 mph. How far has it travelled in
a 4 hours b 7 hours?

8 A forester plants 325 trees on a hectare of land. How many trees will he need to cover a 4.2 hectares b 12.6 hectares?

9 A tanker holds 27.5 m^3 of milk. Find the volume of milk delivered to the depot when a 4 tankers b 15 tankers have unloaded there.

10 Three hikers carry haversacks with masses of 20.6 kg, 18.9 kg, and 24.5 kg. What is the total mass which they carry?

11 a The three sides of a triangle are 12.8 cm, 14.6 cm and 8.7 cm long. Find the perimeter of the triangle.
b Another triangle has sides 4.24 m, 2.85 m and 3.08 m long. Find its perimeter.

12 Find the perimeter of a rectangle
a 3.6 cm long and 2.7 cm wide b 27.8 cm long and 15.4 cm wide.

13 A cyclist averages a speed of 21.8 km/h. How far did he travel if the journey took him
a 3 hours b 6.5 hours?

14 A couple buy a 6.5 m length of carpet costing £5·46 per metre.
a How much do they spend? b If they had chosen a cheaper carpet at £4·82 per metre, what would it have cost them?

15 If there are 72.5 kilocalories in one slice of bread, how many kilocalories will there be in
a 4.5 slices b a loaf of 32 slices?

16 A gas bill shows that the price of each therm of gas is 27·5 pence. Find the cost of gas if
a 324 therms b 253.6 therms are used. Answer in £s.

17 From a ball of string four pieces can be cut which are 2 m, 1.73 m, 0.87 m and 5.6 m long. Find the total length of string in the ball.

18 A basket of groceries holds 5 kg of apples at £0·75 per kg, a 3 kg joint of beef at £2·65 per kg, 3 kg of sugar at £0·28 per kg, and 4 pints of milk at £0·19 per pint. Find the total cost of all the groceries.

Decimal problems

19 A hotel charges per person £19·50 for bed and breakfast and an evening meal is £4·25 extra. How much will it cost two people to stay for 15 days at this hotel if they have bed, breakfast and evening meal?

20 How many times does the turntable of a record-player go round when it plays a record which lasts for
a 3.5 minutes at 45 r.p.m. b 27.0 minutes at 33.3 r.p.m.?

Part 2

1 During the five working days of a week, Mr Horbury spent the following amounts on petrol, £3·50, £2·47, £8·64, £12·72 and £1·79. How much did he spend altogether on petrol during this week?

2 Four houses are built on a plot of land so that the areas they occupy are 0.42 hectares, 0.34 hectares, 0.38 hectares and 0.46 hectares. What is the total area of this land?

3 If $\widehat{AOB} = 24.5°$, $\widehat{BOC} = 52°$ and $\widehat{COD} = 31.5°$, what is the size of angle AOD?

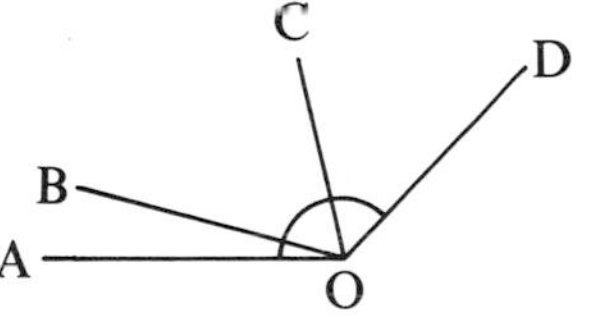

4 Four bottles holding 2 litres, 1.75 litres, 0.625 litres and 1.375 litres of liquid have their contents poured into one large container. How much liquid is in this container?

5 One kilogram of plums costs £0·84. How much will
a 5 kg b 14 kg c 3.6 kg d 12.4 kg cost?

6 The petrol consumption of a new car is said to be 9.4 miles per litre. How far should it travel on
a 6 litres b 13 litres c 2.5 litres d 23.5 litres of petrol?

7 The boat sailing across the North Sea from Bergen, in Norway, to Newcastle averages a speed of 22.6 nautical miles per hour.
a If the journey takes 20 hours, find the distance from Bergen to Newcastle.
b If the boat stops at Stavanger, the sailing time is 22.5 hours. Find the distance the boat has to travel.

8 Mrs Jackson buys 2 kg of sugar, 4.5 kg of potatoes, 0.125 kg of tea, 0.65 kg of cheese, 1 kg of onions and 0.375 kg of tomatoes.
a What is the total mass in her shopping bag?
b If her bag has a mass of 1.8 kg, what mass does she carry home?

9 The bedrooms of a house have windows with the following areas: 2.4 m^2, 1.25 m^2, 1.45 m^2, 2.4 m^2.
a What is the total window area in the bedrooms?
b If double-glazing costs £34 per m^2 of windows, find the cost of double-glazing these four windows.

10 Five resistors in series in an electrical circuit have the following voltage drops across them 42.7, 126.5, 24.8, 2.45, 8.35 volts.
Find the total voltage drop across all five resistors.

11 Mrs Netherton buys four lengths of material at the local market, each costing £1·46 per metre. The lengths are 3.25 m, 6 m, 4.5 m and 1.75 m. Find
a the total length she buys b the total price she pays.

Decimal problems

12 If Mr Ryton uses his car on business, he can claim 16·5 pence for every mile he travels. On the five days of one week he travelled 26.5 miles, 14.7 miles, 12 miles, 36 miles and 32.8 miles.
- a What is the total distance he travelled?
- b How much can he claim?

13 A car owner bought a tin of oil and used 2.4 litres of it immediately. He then filled a 4-litre container and had 1.1 litres left in the tin.
- a How many litres of oil did he buy?
- b If one litre costs £0·96, how much did the tin of oil cost?

14 Three sisters go on a 24-mile sponsored walk and they all complete the full distance. Ann was sponsored for £1·23 per mile, Karen for £2·04 per mile and Jane for £1·65 per mile.
- a How much per mile were they sponsored altogether?
- b How much sponsor money did they collect altogether after the walk?

15 Five sacks of potatoes with masses of 24 kg, 22 kg, 19.5 kg, 26.5 kg and 25 kg are delivered to a shop. The shopkeeper will be selling them at 18 pence per kg. Find
- a the total mass of potatoes delivered to the shop
- b the money the shopkeeper will receive when he has sold them all.

16 A grocer bought some goods to sell in his shop. He bought 4 boxes of dog food at £4·56 per box, 12 cartons of sugar at £3·62 per carton, 25 packs of razor-blades at £1·47 per pack and 8 crates of fruit drinks at £2·36 per crate. How much did he spend altogether?

17 A coach travelling between several towns carries passengers paying different rates. 3 passengers paid £3·54 each, 6 passengers paid £1·84, 23 passengers paid £5·78 and 14 passengers paid £4·29. Find the total amount paid in fares on this journey.

18 When Mr Felton drives to work, he sometimes gives a lift to four of his workmates and they each make a contribution to the petrol. In March, Bill had 12 lifts and he pays 25p per trip, John had 24 lifts and he pays 14p per trip, Malcolm had 8 lifts at 18p per trip, and Tony had 36 lifts at 22p per trip. How much did Mr Felton receive altogether?

19 In the country of Bodemia there are 4.25 million men, 4.17 million women, 1.2 million girls and 0.9 million boys.
- a How many million people are there in Bodemia?
- b For every one person in Bodemia there are 14 pigs. How many million pigs are there?

20 Mrs Large goes by air on her holiday. She herself has a mass of 85.5 kg and the clothes she wears have a mass of 3.5 kg. She takes her full luggage allowance of 20 kg and also 12.5 kg of excess luggage.
- a How heavy is Mrs Large with all her belongings?
- b The airline charges £3·68 for each kg of excess luggage. How much will they charge Mrs Large?

Part 3

1 In one pocket a boy has £1·56. In his other pocket he has £2·72. How much has he altogether?

2 June has £1·17, Carol has £3·27 and Lucy has £3·94. How much have the three girls altogether?

Decimal problems

3 One kilogram of potatoes costs £0·21.
How much does a 4 kg b 12 kg cost?

4 A boy earns £0·82 in one hour.
How much does he earn in a 6 hours b 14 hours?

5 Mr Clough travels 42 miles every day to and from work.
a How far does he travel in a working week of 5 days?
b How far does he travel in 12 working weeks?

6 A gardener plants 32 lettuces to one square metre. How many plants will he put into a 7 square metres b 21 square metres?

7 On average, John's stride is 85 cm.
How far will he move when he takes a 5 strides b 24 strides?

8 Water flows out of a tap at a rate of 62.7 litres per minute.
How much flows out of the tap in
a 3 minutes b 15 minutes?

9 A train travels at a steady speed of 96.8 kilometres per hour.
How far will it travel at this speed in
a 2 hours b 2.5 hours?

10 Curtain material costs £4·16 per metre. How much would
a 3 metres b 3.5 metres of this material cost?

11 David has a mass of 39.2 kg and his sister 35.8 kg. Their parents have a mass of 76.3 kg and 57.3 kg.
a What is the total mass of this family?
b If 1 kg = 2.2 lb, what is their total mass in lb?

12 You go on a holiday to France and you change £1 for 11.26 francs. How many francs will you get for a £5 b £16 c £62·50?

13 One tin of baked beans has a mass of 0.58 kg. 48 tins are packed into a cardboard box. The box has a mass of 0.25 kg. Find the total mass of a full box.

14 The flight from London to Nairobi takes 13 hours. If the aeroplane averages a speed of 341 miles per hour, find the distance between London and Nairobi.

15 A young girl decorates a table-cover which is 3.25 metres by 2.75 metres. She decides to sew a lace frill round the edge. How long will this lace edge have to be?

16 A boiled egg takes 4.5 minutes before it is hard. If you boil three eggs in the same pan together, how long have you to wait before they are hard?

17 A woman pays £12·65 each month for two years to buy a suite of furniture. How much does she pay altogether? If the cash price was £284·45, how much extra has she had to pay?

18 A British Rail diesel engine has a mass of 845 tonnes. It can pull a train 4.6 times its own mass. How heavy are both engine and train together?

19 A man drives by car from Southampton to Liverpool. On ordinary roads his average speed is 38 miles per hour and this takes him 2.3 hours. On motorways his average speed is 58 miles per hour and this part of the journey takes him 2.3 hours. Find the distance between these two cities.

20 If a man pays his debts promptly he will get a reduction of 8 pence in each £1. If he owes £126·25, what would be his total reduction for prompt payment?

21 The concrete base for a garage is 12.65 metres by 8.72 metres. Find the area of this rectangular base to the nearest square metre.

22 How many seconds are there in one year?

Decimal problems

Subtraction and division

Part 1

1 David had £8·64 before he spent £2·38. How much has he left?

2 Linda saved £6·39 and decided to buy a record costing £4·54.
 a How much would she have left after buying the record?
 b If she also spent £0·67 in bus fares going to the shop, how much would be left?

3 A young couple bought a washing machine for £214·36 and paid £58·50 as a deposit. How much did they still have to pay.

4 If $\widehat{XOZ} = 135.4°$ and $\widehat{XOY} = 97.2°$, what is the size of angle YOZ?

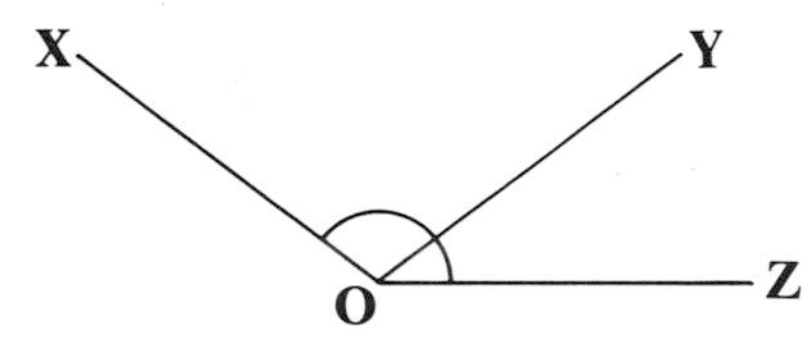

5 A beaker of water from a hot tap has a temperature of 61.7°C. During the next few minutes, the temperature falls by 12.9°. What is the new temperature?

6 a If six litres of petrol cost £2·28, find the cost of one litre.
 b If eight litres of a certain oil cost £3·44, find the cost of one litre.

7 A small firm buys four new cars for a total of £21 474. If all the cars are the same model, find the cost of one.

8 A piece of elastic 27.6 m long is cut into eight equal pieces.
 a Find the length of each piece.
 b If one of these pieces is now cut into five equal lengths, how long will each length be?

9 A sheet of cardboard with an area of 2.87 m^2 is cut into five equal pieces.
 a Find the area of each piece.
 b If one of these pieces is now cut into four equal pieces, what is the area of one of these smaller pieces?

10 A housewife has a 2.5 kg bag of flour.
 a A recipe requires 0.75 kg of flour. How much will she have left?
 b She then makes several cakes using 1.125 kg of flour. What mass of flour remains?

11 a Mrs Emley pays for £3·60 worth of groceries with a £5 note. How much change will she receive?
 b Her husband buys a new car tyre costing £23·45 with three £10 notes. What change does he receive?
 c Their daughter spends £12·64 on a new pair of shoes. What change will she get from three £5 notes?

12 a Two angles add up to 90°. If one angle is 63.5°, find the other angle.
 b Two angles add up to 180°. If one angle is 72.4°, find the other angle.

13 A vat of dye holds 15.5 litres but a quarter of the full vat leaks away through a hole in the bottom. How many litres are lost?

14 a Mr Geldby pays one sixth of his salary in tax. If he earns £513·84 in a month, how much does he pay in tax?
 b His wife pays one fifth of her salary as tax. If she earns £304·25, how much tax does she pay?

Decimal problems

15 May Thomas spent £12·70 last Saturday. One tenth of this was on bus fares.
  a How much did she spend on bus fares?
  b How much was spent in other ways?

16 A roll of wallpaper has an area of 7.8 m^2. When John Howe wallpapers he wastes one twelfth of it.
  a What area of paper is wasted?
  b What area of paper is used?

17 A greengrocer opens a full 60 kg bag of potatoes and at the end of the day only one eighth ($\frac{1}{8}$) of the potatoes are left.
  a What mass of potatoes was left?
  b How many kilograms had been sold?

18 A packet of eight bars of chocolate costs £2·76. The shopkeeper opens the packet and sells one bar to three boys.
  a How much does one bar of chocolate cost?
  b If the three boys share the one bar equally, how much should they each pay for their chocolate?

19 Three boys share a 33 kg sack of apples. Garry takes one third ($\frac{1}{3}$) of them, Richard takes one quarter ($\frac{1}{4}$) of them and Adrian has the rest. Find the mass of apples for each boy.

20 A X B The distance from A to B is 10.8 cm.
  a The distance XB is one quarter of the distance AB. Find the distance XB.
  b Find the distance AX.
  c If AX is divided into six equal parts, find the length of each part.

Part 2

1 Today is Anita Jackson's birthday and she has a mass of 51.4 kg. On her last birthday she had a mass of 46.1 kg. How many kg has she gained this year?

2 Mr North had 186.5 m^2 of lawn in his garden, but now he grows more vegetables and he has only 98 m^2 left. How much lawn has he taken away?

3 The Parish of Grubthorpe has an area of 28 km^2 of which 6.5 km^2 is built-up. What area of the parish is not built-up?

4 a A coat costing £58 is reduced in a sale to £45·60. How much is this reduction?
  b In the same sale, another coat is priced at £37·40. If its original price was £51, what is this reduction?

5 a If 4 kg of stewing steak cost £10·08, find the cost of one kilogram.
  b If 3 kg of fish cost £5·79, find the cost of one kilogram.
  c If 8 kg of sprouts cost £5·84, find the cost of one kilogram.

6 a A car takes 5 hours to travel the 203 miles to Plymouth. Find its average speed in miles per hour.
  b A train makes the same journey in 4 hours. What is its average speed in miles per hour?

7 To spend a day at the coast, a boy cycles 52.5 miles and gets there in 5 hours. Find his average speed in miles per hour.

8 10.8 m^2 of sheet metal are used to make tin cans and one ninth of the metal is wasted in the process. Find
  a the area of wasted metal b the area of metal used.

Decimal problems

9 The winner of an 8-lap race had a time of 12.5 minutes. The time for the last lap was one tenth of the total time. Find
a the time for the last lap b the time for the remaining 7 laps.

10 One twelfth of a fruit squash is concentrated fruit juice and the rest is water. In a large container holding 16.2 litres of fruit squash, what is the quantity used of
a concentrated fruit juice b water?

11 A boy buys a bicycle for £124·35 by paying £37·43 immediately and the remainder in four equal weekly instalments.
a How much was left to pay after the first payment?
b What amount did he have to pay each week?

12 A man buys 2.3 litres of turpentine and uses 0.35 litres straight away. The remainder of the turpentine will just fill six identical bottles.
a How much had he left after using the 0.35 litres?
b How much will each bottle hold?

13 A handyman buys 4.5 m^2 of plywood and trims it to size wasting 0.22 m^2. He cuts the remainder into eight pieces of equal size.
a How much plywood was left after he had trimmed it?
b What was the area of one of the eight pieces?

14 A box holds seven large packets of Sugared Weeties and the total mass is 9 kg. The box alone has a mass of 0.25 kg. Find the mass of
a the seven packets b one packet.

15 Three skirts, needing equal amounts of cloth, are to be made from 5.7 m^2 of material. Mrs Andrews cuts off a length for the first skirt and after making it there is 0.16 m^2 of waste.
a What area of material is cut off for each skirt?
b How much material actually went into the first skirt?

16 a Mr Clough's car engine needs 4 litres of oil. He has a bottle holding 2.65 litres. How much more oil does he need?
b If he buys a 3-litre tin of oil at the local supermarket, how much will he have left after filling his engine?

17 A right angle is divided into four smaller angles, one of which is 18.6°. The other three angles are all equal. Find the size of these equal angles.

18 a The first payment on a car costing £3565 is £602·50. How much is still left to pay?
b If the second payment is £426·78, what amount is now left?
c The remainder is paid in 12 equal instalments. How much is each of these instalments?

19 XZ is 10.2 cm long and YZ is one quarter of XZ.
a Find the length of YZ.
b Find the length of XY.
c YZ is divided into six equal parts. What is the length of one of these parts?

X Y Z
10.2 cm

20 Mr and Mrs Spedding buy 20.4 m^2 of carpet for two rooms of their house. They cut off one fifth of the piece and fit it in the smaller room, wasting 0.29 m^2 of carpet.
a How much carpet do they cut off for the smaller room?
b What is the area of the smaller room?
c How much carpet is left for the other room?
d If 0.35 m^2 is wasted when fitting the second room, what is the area of its floor?

Decimal problems

Part 3

1 Tracy has £4·62. If she spends £2·91, how much has she left?

2 On a touring holiday, a family planned to travel 1500 miles. In fact they travelled 2365 miles. How much further did they go than they had intended?

3 The room temperature is 21.5°C. When the central heating is turned off, the temperature falls 7.4°C. What is the new temperature?

4 Mrs Brigshaw had a mass of 88.5 kg but then she went on a diet and lost 19.4 kg. What is her mass now?

5 A jar full of jam has a mass of 502 grams. When it is empty, the jar alone has a mass of 48 grams. What is the mass of the jam?

6 Six new delivery vans cost £24 792. How much does one van cost?

7 A car travels 108 miles on 8 litres. How far does it travel on one litre?

8 A cricketer scores 294 runs in 7 innings. What is his average number of runs for each innings?

9 12 kilograms of potatoes cost £3·84. How much does 1 kg cost?

10 A man earns £15·68 in an 8-hour day. How much does he earn in one hour?

11 A piece of curtain tape is 8.61 metres long. If one third is cut off, how many metres are left?

12 A full water tub holds 146 litres. If $\frac{3}{4}$ of the water is used, how many litres are left?

13 A car starts a journey with a full tank of petrol. At the end of the trip the tank is three-fifths full, and the car has used 18 litres of petrol. How many litres does a full tank hold?

14 A piece of string is 8.64 metres long. One third of it is used and then a second piece 1.29 metres long is cut off. How much string is left?

15 If £13 are changed for 146.25 French francs, how many francs will £1 be worth?

16 A large bakery buys 54 tonnes of grain for £4320. How much is 1 tonne worth?

17 1.2 metres of curtain material costs £5·52. What is the price for 1 metre?

18 A car has 2.4 litres of petrol in its tank, and on this petrol it runs for 39.6 km. How many kilometres does it travel on one litre.

19 The fastest express train from Leeds to London takes only 2.5 hours. If it travels a distance of 197.5 miles, find its average speed in miles per hour.

20 A stopping train takes 1.5 hours to go from Manchester to Leeds. The distance between these cities is 42.0 miles. Find the average speed of the train in miles per hour.

21 A bath takes 8.4 minutes to fill, and when it is full it holds 273 litres. How many litres are coming out of the tap every minute?

22 A farmer uses 73.5 kg of fertiliser on 14 hectares of land. How much fertiliser will he need for

a 1 hectare   b 23 hectares?

A mixture

1 A hiker packs 4 kg of clothes, 3.5 kg of food, a 4.25-kg tent and 1.8 kg of other equipment into her haversack which itself has a mass of 0.74 kg. What is the total mass she carries on her back?

Decimal problems

2 Two boys go on a week's cycling holiday. In the first six successive days they travel 74 miles, 85 miles, 72 miles, 78 miles, 65 miles and 82 miles.
  a What is the total mileage they travel during these six days?
  b If they covered 517 miles altogether, how far did they cycle on the last day?
  c If the cost of the week's holiday is £23·94, how much is this per day.

3 A man has a rectangular garden 15 metres by 6.5 metres.
  a How long will a fence need to be to go right round it?
  b If the fence he chooses costs £4·55 per metre, how much will it cost?

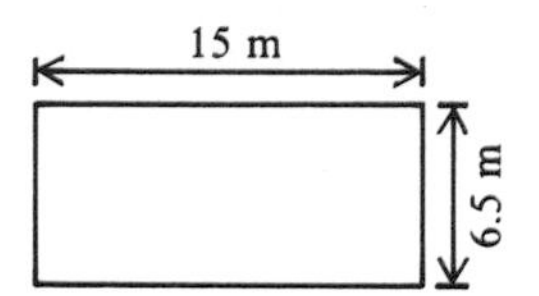

4 A man leaves £5021 in his will to be shared equally between his three daughters and one son. How much do they each receive?

5 An apple tree in an orchard needs 9 square metres of ground.
  a If the orchard covers 2340 square metres, how many trees can be planted?
  b If one tenth of the trees die on planting, (i) how many die, (ii) how many survive?

6 a Find $\frac{1}{5}$ of £27·40. b Find $\frac{3}{5}$ of £27·40.

7 a Find $\frac{1}{8}$ of 139.2 metres. b Find $\frac{5}{8}$ of 139.2 metres.

8 Mr Thwaites has £145·62 in his bank account. He pays into this account a further sum of £47·95 but then he spends two amounts of £28·42 and £34·56. How much is now left in Mr Thwaites' account?

9 If 1 cm^3 of copper has a mass of 8.8 grams, find the mass of
  a 10 cm^3 b 23.5 cm^3 c one litre of copper.

10 A garden lawn is a rectangle 12.5 m by 8.4 m. The gardener needs 25 grams of grass seed to each square metre, and grass seed costs £2·40 per kilogram. Find
  a the area of the lawn
  b the grass seed needed to plant it
  c the cost of the grass seed.

11 Two articles with masses of 2 kg 743 g and 1 kg 376 g are placed in a box. When a third article is added the total mass is 5 kg. Find the mass of the third article.

12 In the four weeks of February, Jackie saves £28·74. In the first week she saved £5·32, in the second week she saved £12 and in the third week she saved £4·60. How much did she save in the fourth week?

13 Mr Clauson earns a basic £95·32 per week.
  a If he has six weeks unpaid holiday in a year, what is his total basic wage for the year?
  b If he also earns £537·68 in overtime during the year, what is his total yearly income?

14 One kilogram is approximately 2.2 pounds. How many pounds approximately are there in a half a kilogram b 5 kg c 23.5 kg?

15 One litre is approximately 1.76 pints. How many pints approximately are there in
  a half a litre b five litres c 45 litres?
  d If there are 8 pints in a gallon, approximately how many gallons are there in 45 litres?

Decimal problems

16 How many children can be given 24p to spend from a total sum of £30?

17 Beech saplings are planted 30 cm apart to make a hedge 45 metres long. How many saplings are needed?

18 How many pieces of string can be cut from a 5-metre length if each piece is 12 cm long? What length will be left?

19 A car engine needs 4.5 litres of oil. How many of these engines can have their oil changed from a 58.5-litre drum?

20 June is making a dress from 3 metres of cloth which costs £1·24 per metre.
- a How much does the cloth cost?
- b If she has a 25 cm strip left over, what length did she use?
- c What is the cost of this 25 cm strip?

21 The outside diameter of a hollow pipe is 7.3 cm, and its inside diameter is 6.7 cm. Find the thickness of the metal pipe in millimetres.

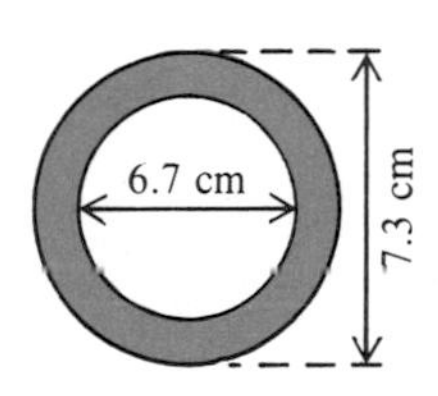

22 A beaker, full of water, has a mass of 102.5 grams. The beaker alone has a mass of 34.9 grams. Find
- a the mass of water which fills the beaker
- b the total mass when the beaker is half full of water.

23 Another beaker, full of water, has a mass of 136.7 grams, and the beaker itself has a mass of 41.2 grams.
- a Find the mass of water which fills the beaker.
- b The beaker is now filled with a liquid which is 2.5 times heavier than water. What is the total mass of the beaker and the liquid.

24 A photograph 12 cm by 7.5 cm is mounted on white card to leave a margin 25 mm wide all around it. Find
- a the area of the photograph by itself
- b the area of the photograph and card when mounted
- c the area of the white margin
- d the perimeter of the photograph
- e the perimeter of the white card.

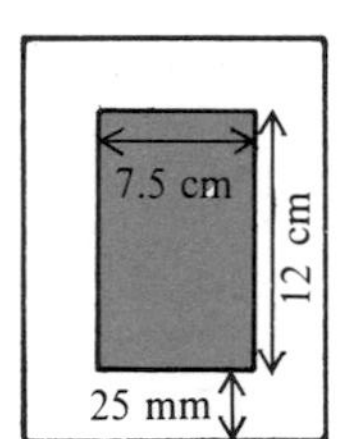

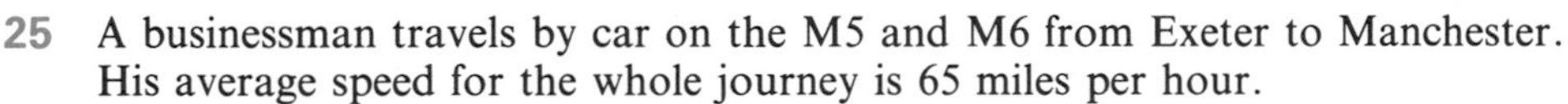

25 A businessman travels by car on the M5 and M6 from Exeter to Manchester. His average speed for the whole journey is 65 miles per hour.
- a How far has he gone after travelling for 2 hours 30 minutes?
- b If Exeter to Manchester is 247 miles, how much further has he to go?
- c How much longer will his journey take him?

Multiplying and dividing by 10, 100 etc.

1 Copy this table. Write in your answers to the multiplications.

		1000	100	10	1	$\frac{1}{10}$	$\frac{1}{100}$	$\frac{1}{1000}$
a	4.35 × 10 =				.			
b	8.16 × 10 =				.			
c	12.8 × 10 =				.			
d	25.7 × 10 =				.			
e	167.0 × 10 =				.			
f	355.0 × 10 =				.			
g	0.12 × 10 =				.			
h	0.85 × 10 =				.			
i	0.04 × 10 =				.			
j	1.27 × 100 =				.			
k	5.34 × 100 =				.			
l	16.87 × 100 =				.			
m	28.5 × 100 =				.			
n	0.67 × 100 =				.			
o	0.39 × 100 =				.			
p	0.052 × 100 =				.			

2 Write the answers only.

a 8.61 × 10
b 7.05 × 10
c 0.67 × 10
d 0.84 × 10
e 12.8 × 10
f 47.3 × 10
g 5.67 × 100
h 0.843 × 100
i 0.605 × 100
j 0.733 × 1000
k 0.105 × 1000
l 1.784 × 1000
m 2.65 × 1000
n 0.06 × 1000
o 0.8 × 100
p 0.05 × 10

3 Each box □ indicates a missing number. Write the missing numbers.

a 7.42 × □ = 74.2
b 0.613 × □ = 6.13
c 0.613 × □ = 61.3
d 1.35 × □ = 135
e 1.35 × □ = 13.5
f 0.07 × □ = 0.7
g 0.07 × □ = 7
h 0.07 × □ = 70
i □ × 6.8 = 68
j □ × 7.8 = 780
k □ × 0.15 = 15
l □ × 0.27 = 2.7
m □ × 0.003 = 0.3
n □ × 0.005 = 5.0
o □ × 3.0 = 300
p □ × 0.6 = 600

Multiplying and dividing by 10, 100 etc.

4 Write the answers only.

- **a** If one nail has a mass of 3.65 grams, what will be the mass of 10 nails?
- **b** A piece of string is 1.25 metres long. Another piece of string is 10 times longer. What is its length?
- **c** A car averages 9.25 miles on one litre of petrol. How far will it travel on 10 litres?
- **d** The winning time of the first race at an athletics meeting is 12.4 seconds. What was the winning time of the second race, if it took 10 times longer?
- **e** One box of chocolates costs £3·60, how much will 10 cost?
- **f** If one pin has a mass of 0.34 grams, what is the mass of 10 pins?
- **g** If one postage stamp has an area of 3.75 cm^2, what is the area of 100 similar stamps?
- **h** One page of a book is 0.015 cm thick. How thick will 100 pages be?
- **i** £8·56 is the price of one square metre of carpet. How much will 100 m^2 cost?
- **j** One ball-bearing has a mass of 0.023 kg. What will be the mass of 1000 ball-bearings?
- **k** A small bottle holds 0.075 litres of liquid. How many litres will fill another bottle if it will hold 1000 times as much?
- **l** Jim Smith walked 8.5 km today. Tomorrow he will travel 100 times further by train. How far will he go by rail?

5 Copy this table. Write in your answers.

		1000	100	10	1	$\frac{1}{10}$	$\frac{1}{100}$	$\frac{1}{1000}$
a	35.2 ÷ 10 =				.			
b	41.6 ÷ 10 =				.			
c	128.5 ÷ 10 =				.			
d	173.0 ÷ 10 =				.			
e	3654 ÷ 10 =				.			
f	8402 ÷ 10 =				.			
g	14.8 ÷ 10 =				.			
h	2.4 ÷ 10 =				.			
i	3.8 ÷ 10 =				.			
j	0.25 ÷ 10 =				.			
k	264.5 ÷ 100 =				.			
l	379 ÷ 100 =				.			
m	18.5 ÷ 100 =				.			
n	30.7 ÷ 100 =				.			
o	2.9 ÷ 100 =				.			
p	1.7 ÷ 100 =				.			

Multiplying and dividing by 10, 100 etc.

6 Write the answers only.

a	78.1 ÷ 10	**i**	376.2 ÷ 100
b	30.4 ÷ 10	**j**	18.5 ÷ 100
c	2.77 ÷ 10	**k**	20.7 ÷ 100
d	5.19 ÷ 10	**l**	479.8 ÷ 1000
e	237.5 ÷ 10	**m**	6790 ÷ 1000
f	489 ÷ 10	**n**	8600 ÷ 1000
g	6774 ÷ 10	**o**	2.6 ÷ 100
h	8940 ÷ 10	**p**	1.7 ÷ 100

7 Each box □ indicates a missing number. Write the missing numbers.

a	89.5 ÷ □ = 8.95	**i**	3.67 ÷ □ = 0.367
b	14.7 ÷ □ = 1.47	**j**	14.2 ÷ □ = 0.142
c	14.7 ÷ □ = 0.147	**k**	40 ÷ □ = 0.4
d	285.1 ÷ □ = 2.851	**l**	75.2 ÷ □ = 0.752
e	678 ÷ □ = 6.78	**m**	75.2 ÷ □ = 0.0752
f	3592 ÷ □ = 3.592	**n**	86.5 ÷ □ = 0.0865
g	6080 ÷ □ = 6.08	**o**	3.6 ÷ □ = 0.036
h	1.7 ÷ □ = 0.17	**p**	1 ÷ □ = 0.001

8 Write the answers only.

a If 10 bottles of the same size hold 14.5 litres altogether, how much will one bottle hold?

b Ten books of the same size are placed in a pile. If the pile is 32.4 cm high, how thick is one book?

c 10 similar plates have a total mass of 565 grams. What is the mass of one plate?

d In 10 hours a cyclist travels 246 km. How far does he average in one hour?

e 100 small ball-bearings together have a total mass of 92.5 grams. What is the mass of one?

f 100 sheets of cardboard are stacked in a pile 18.6 cm high. How thick is one sheet of cardboard?

g A room is fitted with 100 m^2 of carpet costing £845. How much does 1 m^2 cost?

h A book is made from 100 sheets of paper. If the book is 2.6 cm thick, how thick is one sheet of paper?

i (i) How many cm are in 1 metre?
(ii) If a steel bar 1 metre long has a mass of 2.5 kg, what is the mass of 1 cm of the bar?

j 1000 floor tiles will cover an area of 176.5 m^2. What is the area of one tile?

k 1000 postage stamps have a total mass of 16.2 grams. What is the mass of one stamp?

l (i) How many cm^3 are in 1 litre?
(ii) If 1 litre of paraffin has a mass of 970 grams, what is the mass of 1 cm^3?

m (i) How many grams are in 1 kg?
(ii) If 1 kg of tin foil covers an area of 130 m^2, what area does 1 g cover?

Multiplying and dividing by 10, 100 etc.

9 Write the answers only.

a 6.75×10

b $6.75 \div 10$

c 21.3×10

d $\frac{21.3}{10}$

e 8.45×100

f $\frac{16.8}{100}$

g $\frac{10.8}{10}$

h 0.0065×1000

i 0.125×100

j $\frac{68.9}{100}$

k 0.425×1000

l $\frac{62.5}{1000}$

m 10×0.07

n 100×0.4

o $\frac{4650}{1000}$

p 1000×0.001

10 Write the answers only.

a 6.4×10

b 6.4×100

c $\frac{2.8}{10}$

d $\frac{2.8}{100}$

e 0.16×1000

f 0.08×1000

g $\frac{62.5}{100}$

h $\frac{62.5}{1000}$

i 5.02×100

j 5.02×1000

k $\frac{40}{100}$

l $\frac{38}{1000}$

m $0.752 \times 10\,000$

n $0.015 \times 10\,000$

o $\frac{1250}{10\,000}$

p $\frac{3}{100}$

Change of units

1 Which units of length would you use to measure the following?

km **m** **cm** **mm**

- a the distance from Inverness to Penzance
- b the length of a garden path
- c the height of a television screen
- d the thickness of a sheet of cardboard
- e the length of a bus
- f the width of an exercise book
- g the distance you travel to school
- h the thickness of a sheet of glass

2 Which units of mass would you use to weigh the following?

kg **g**

- a a week's supply of potatoes
- b one plum
- c a bag of cement
- d an egg
- e your own mass
- f a tin of peas

3 Which units of volume would you use to measure the following?

litres **cm³**

- a a tin of motor oil
- b a dose of medicine
- c a tank of petrol
- d a bottle of perfume
- e a bottle of milk
- f a glass of wine

4 Write these amounts in pence.

a	£2·78	b	£9·42	c	£3·04	d	£16·75
e	£26·10	f	£87·45	g	£154·72	h	£215·50
i	£0·75	j	£0·14	k	£0·07	l	£18
m	£246	n	£2375				

5 Write these amounts in £.

a	300p	b	365p	c	897p	d	1240p
e	2780p	f	5080p	g	9500p	h	14000p
i	37650p	j	275p	k	86p	l	14p
m	8p	n	2p				

Change of units

6 Write these masses in kilograms.

a	2 kg 645 g	b	5 kg 167 g	c	7 kg 875 g
d	1 kg 160 g	e	3 kg 87 g	f	5 kg 45 g
g	0 kg 70 g	h	2 kg 50 g	i	3 kg 8 g
j	1 kg 5 g	k	3 kg 2 g	l	0 kg 1 g

7 Write these volumes in litres.

a	4 litres 675 cm^3	b	8 litres 270 cm^3	c	2 litres 126 cm^3
d	1 litre 940 cm^3	e	3 litres 85 cm^3	f	12 litres 46 cm^3
g	0 litres 10 cm^3	h	9 litres 70 cm^3	i	4 litres 5 cm^3
j	2 litres 4 cm^3	k	0 litres 12 cm^3	l	0 litres 1 cm^3

8 Write these lengths in kilometres.

a	3 km 675 m	b	8 km 965 m	c	7 km 125 m
d	1 km 500 m	e	1 km 320 m	f	24 km 800 m
g	2 km 75 m	h	0 km 82 m	i	1 km 40 m
j	0 km 8 m	k	1 km 4 m	l	0 km 1 m

9 Write these lengths in metres.

a	2 m 85 cm	b	6 m 14 cm	c	8 m 90 cm
d	1 m 96 cm	e	0 m 34 cm	f	2 m 8 cm
g	1 m 7 cm	h	0 m 6 cm	i	14 m 14 cm
j	2 m 2 cm	k	13 m 1 cm	l	0 m 1 cm

10 Write these lengths in centimetres.

a	8 cm 6 mm	b	9 cm 4 mm	c	12 cm 2 mm
d	94 cm 8 mm	e	1 m 26 cm 7 mm	f	2 m 97 cm 4 mm
g	6 m 42 cm 1 mm	h	3 m 9 cm 5 mm	i	6 m 7 cm 8 mm
j	6 m 0 cm 5 mm	k	2 m 0 cm 9 mm	l	1 m 0 cm 1 mm

11 Write these masses in kilograms.

a	3000 g	b	6500 g	c	2750 g
d	14800 g	e	28250 g	f	1425 g
g	984 g	h	642 g	i	106 g
j	75 g	k	37 g	l	4 g

12 Write these masses in grams.

a	5 kg	b	8.2 kg	c	4.35 kg
d	7.12 kg	e	3.625 kg	f	0.875 kg
g	0.003 kg	h	0.06 kg	i	0.72 kg
j	0.9 kg	k	0.85 kg	l	0.009 kg

Change of units

13 Write these volumes in litres.

a	6000 cm^3	b	3800 cm^3	c	4750 cm^3
d	25400 cm^3	e	1070 cm^3	f	975 cm^3
g	136 cm^3	h	99 cm^3	i	24 cm^3
j	8 cm^3	k	2 cm^3	l	0.5 cm^3

14 Write these volumes in cm^3.

a	5 litres	b	9.7 litres	c	13.25 litres
d	16.05 litres	e	3.125 litres	f	0.675 litres
g	0.045 litres	h	0.008 litres	i	0.05 litres
j	0.53 litres	k	0.03 litres	l	0.4 litres

15 These lengths are given in either metres or millimetres. Write them in **centimetres**.

a	4 m	b	4.25 m	c	5.75 m	d	12.84 m
e	18.75 m	f	26.74 m	g	0.46 m	h	0.27 m
i	0.99 m	j	0.03 m	k	2 m	l	14 m
m	20 mm	n	25 mm	o	15 mm	p	87 mm
q	99 mm	r	101 mm	s	769 mm	t	846 mm
u	1000 mm	v	28 mm	w	8 mm	x	2 mm

16 These lengths are given in either kilometres or centimetres. Write them in **metres**.

a	2 km	b	2.4 km	c	2.46 km	d	2.462 km
e	8 km	f	8.75 km	g	1.03 km	h	6.125 km
i	14 km	j	12.5 km	k	0.75 km	l	0.8 km
m	300 cm	n	350 cm	o	270 cm	p	125 cm
q	999 cm	r	1006 cm	s	3625 cm	t	104 cm
u	99 cm	v	34 cm	w	8 cm	x	3 cm

17 This table gives measurements in mixed units. Write each measurement in **a** metres only **b** cm only **c** mm only.

Example 2 m 48 cm 2 mm = 2.482 m = 248.2 cm = 2482 mm

	km	m	cm	mm
a		3	26	8
b		4	17	2
c		27	91	4
d		31	6	5
e		324	9	7
f		607	7	5
g	2	154	80	2
h	3	67	9	5
i		3	24	
j		1	6	

Change of units

18 Work the following. Give your answers in £s.

a £1·64 + £2·37 + 68p + 84p
b 95p + £12·42 + £3·80 + 127p
c £32·78 + 87p + £12·05 + 247p
d 64p + £0·95 + 212p + £1·62
e £2·87 − 59p
f £8·45 − 253p
g 409p − £0·37
h £12·64 − 328p

19 Work the following. Give your answers in kilograms.

a 3.825 kg + 2.424 kg + 875 g
b 0.736 kg + 1.24 kg + 127 g
c 0.8 kg + 3.74 kg + 370 g + 85 g
d 489 g + 0.075 kg + 1.6 kg + 34 g
e 3.125 kg − 855 g
f 4975 g − 1.62 kg
g 2.85 kg − 325 g
h 1.4 kg − 270 g

20 Work the following. Give your answers in litres.

a 1.692 litres + 0.75 litres + 456 cm^3
b 0.875 litres + 0.9 litres + 260 cm^3
c 2.4 litres + 575 cm^3 + 0.25 litres + 3640 cm^3
d 0.37 litres + 284 cm^3 + 1.6 litres + 2066 cm^3
e 8.625 litres − 875 cm^3
f 2140 cm^3 − 1.7 litres
g 4.6 litres − 370 cm^3
h 0.85 litres − 146 cm^3

21 Work the following. Give your answers in centimetres.

a 12 cm + 4.8 cm + 17 mm
b 37 cm + 65 mm + 5.8 cm
c 84 mm + 5 cm + 2.1 cm
d 0.7 cm + 2 cm + 4 mm
e 8.7 cm − 14 mm
f 32 cm − 45 mm
g 68 mm − 1.9 cm
h 240 mm − 13.2 cm

Change of units

22 Work the following. Give your answers in metres.

- a 8.6 m + 300 cm + 140 cm
- b 12.4 m + 545 cm + 95 cm
- c 2.625 km + 647 m + 0.855 km
- d 0.136 km + 87 m + 500 cm
- e 32.4 m + 0.071 km + 120 cm
- f 6.75 m − 114 cm
- g 3.05 m − 72 cm
- h 408 cm − 1.3 m
- i 1.125 km − 621 m
- j 0.714 km − 109 m

23
- a Three bottles of orange juice hold 2.2 litres, 675 cm^3 and 1200 cm^3. How many litres are there altogether?
- b Three planks of wood are 4.8 cm, 6 mm and 17 mm thick. If they are nailed together, what is their total thickness in cm?
- c 4 cm of rain fell on Monday, 1.6 cm on Tuesday and 3 mm fell on Wednesday. What was the total amount of rain (in mm) which fell on these three days?
- d A metal box has a mass of 625 grams, its lid has a mass of 36 grams and its contents have a mass of 2.4 kg. What is the total mass in kg?
- e A cyclist stops twice on a journey. What was the total distance he went (in km) if the three parts of his trip were 2.7 km, 875 m and 1475 m?
- f Three cardboard boxes stand on top of each other. If their heights are 1.7 m, 136 cm and 94 cm, find their total height in metres.
- g Four containers hold the following volumes of cooking oil: 1.15 litres, 0.625 litres, 375 cm^3 and 520 cm^3. How many litres is this altogether?
- h A 3 m pole is sunk 85 cm into the ground. What height is above ground level?
- i 640 g of flour are taken from a 2 kg-bag. How many grams are left?
- j A ruler 30 cm long has a piece 67 mm long sawn off. How many centimetres are left?
- k 1270 cm^3 are poured from a 2-litre tin of oil. How many cubic centimetres are left?
- l An athlete drops out of the 3000 m race after running 1.25 km. How many metres had he still to run to finish the race?

Revision

Part 1 Moving the decimal point

Work the following.

Exercise 1

1 200×600
2 400×300
3 700×200
4 800×20
5 300×40
6 80×400
7 50×200
8 40×500
9 70×3000
10 400×8000
11 3000×500
12 120×8000
13 700×1100
14 1500×1200
15 6000×4200
16 $170 \times 2100 \times 20$
17 $140 \times 2000 \times 40$
18 1200^2
19 500^3
20 $400^2 \times 70$

Exercise 2

1 0.06×2000
2 0.03×6000
3 0.002×9000
4 0.008×4000
5 0.2×300
6 0.4×700
7 0.5×40
8 0.6×500
9 0.004×7000
10 0.005×8000
11 0.006×200
12 700×0.008
13 0.9×400
14 0.007×5000
15 0.0002×4000
16 0.007×200
17 $0.006 \times 400 \times 200$
18 $0.0008 \times 300 \times 100$
19 $0.007 \times 40 \times 800$
20 $0.04 \times 0.06 \times 5000$

Exercise 3

1 $80 \div 0.02$
2 $90 \div 0.03$
3 $150 \div 0.05$
4 $240 \div 0.6$
5 $12 \div 0.003$
6 $210 \div 0.007$
7 $36 \div 0.012$
8 $48 \div 0.016$
9 $240 \div 0.016$
10 $420 \div 0.14$
11 $7.2 \div 0.36$
12 $4.8 \div 0.24$
13 $5.4 \div 0.027$
14 $0.6 \div 0.003$
15 $0.008 \div 0.002$
16 $0.24 \div 0.004$
17 $0.48 \div 0.016$
18 $0.072 \div 0.012$
19 $0.064 \div 0.0016$
20 $0.144 \div 0.036$

Exercise 4

1 $0.08 \div 200$
2 $0.09 \div 300$
3 $0.6 \div 20$
4 $0.8 \div 40$
5 $0.04 \div 20$
6 $0.16 \div 80$
7 $0.16 \div 800$
8 $0.024 \div 60$
9 $0.18 \div 90$
10 $1.2 \div 300$
11 $5.6 \div 700$
12 $9.3 \div 30$
13 $8.4 \div 200$
14 $0.42 \div 600$
15 $4.8 \div 80$
16 $6.4 \div 50$
17 $0.31 \div 500$
18 $0.75 \div 200$
19 $4.6 \div 30$
20 $1.7 \div 40$

Exercise 5

1 0.02×0.03
2 0.04×0.02
3 0.3×0.03
4 0.02×0.4
5 0.06×0.7
6 0.5×0.09
7 0.3×0.7
8 0.9×0.07
9 0.1×0.4
10 0.3×0.12
11 0.3×1.2
12 0.04×1.2
13 0.006×0.9
14 0.07×0.004
15 0.1×0.006
16 0.07×0.8
17 0.6×0.03
18 0.08×0.9
19 0.5×0.07
20 0.1×0.001

Revision

Part 2 Quick questions

Work the following.

Exercise 1 Multiplication

1 0.6×0.3
2 0.4×0.2
3 1.2×0.4
4 0.02×0.7
5 $(0.6)^2$
6 $(0.04)^2$
7 0.04×8
8 12×0.3
9 2.1×0.2
10 0.2×0.3
11 6×0.4
12 60×0.4
13 600×0.04
14 400×0.7
15 200×0.3
16 3000×0.08
17 0.06×700
18 0.004×5000

Exercise 2 Division

1 $\frac{2.7}{9}$
2 $\frac{2.7}{0.9}$
3 $\frac{3.6}{12}$
4 $\frac{3.6}{1.2}$
5 $\frac{0.48}{0.06}$
6 $\frac{0.49}{0.7}$
7 $\frac{1.2}{0.06}$
8 $\frac{3.2}{0.08}$
9 $\frac{14}{0.2}$
10 $\frac{24}{0.1}$
11 $\frac{1}{0.1}$
12 $\frac{1}{0.5}$
13 $\frac{1}{0.01}$
14 $\frac{0.05}{0.002}$
15 $\frac{0.7}{0.02}$
16 $\frac{14.4}{1.2}$
17 $\frac{1}{0.4}$
18 $\frac{1}{0.02}$

Exercise 3 Mixed

1 0.2×1.2
2 0.6×1.3
3 0.4×0.6
4 0.3×0.7
5 $(0.5)^2$
6 $(1.2)^2$
7 $\frac{1.4}{7}$
8 $\frac{1.4}{0.7}$
9 $\frac{4.8}{0.2}$
10 $\frac{1.6}{8}$
11 $\frac{0.06}{0.03}$
12 $\frac{0.12}{0.04}$
13 $\frac{1.6}{0.08}$
14 $\frac{3.5}{0.07}$
15 $\frac{1}{0.2}$
16 0.2×3.4
17 0.01×150
18 $48 \times (0.1)^2$

Exercise 4 Mixed

1 0.7×0.6
2 0.4×0.8
3 1.5×0.4
4 0.5×0.07
5 1.2×0.03
6 $(0.2)^2$
7 $(0.06)^2$
8 $(0.1)^3$
9 $\frac{2.8}{7}$

Revision

10 $\frac{2.8}{0.7}$

11 $\frac{0.036}{0.009}$

12 $\frac{0.036}{0.09}$

13 $\frac{0.072}{0.9}$

14 $\frac{0.48}{0.6}$

15 $\frac{1}{0.5}$

16 $\frac{2}{0.4}$

17 $\frac{6}{0.02}$

18 $\left(\frac{6}{1.2}\right)^2$

Exercise 5 Multiplication

1 0.1×0.4
2 0.7×0.6
3 0.04×1.2
4 $(0.03)^2$
5 5×0.7
6 0.02×12
7 300×0.7
8 6000×0.03
9 0.002×8000
10 0.0012×300
11 600×0.8
12 60×0.08
13 6×0.008
14 600×8000
15 60×800
16 3000×2000
17 4000×900
18 $(500)^2$

Exercise 6 Division

1 $\frac{4.8}{8}$

2 $\frac{4.8}{0.8}$

3 $\frac{4.8}{0.08}$

4 $\frac{0.09}{0.03}$

5 $\frac{0.4}{0.005}$

6 $\frac{3.6}{0.004}$

7 $\frac{2}{0.1}$

8 $\frac{1}{0.05}$

9 $\frac{1}{0.3}$

10 $\frac{500}{200}$

11 $\frac{50\,000}{200}$

12 $\frac{3600}{9000}$

13 $\frac{40}{500}$

14 $\frac{24}{6000}$

15 $\frac{45\,000}{900}$

16 $\frac{3.5}{700}$

17 $\frac{1.2}{6000}$

18 $\frac{1}{500}$

Exercise 7 Mixed

1 $\frac{1.2}{0.6}$

2 1.2×0.6

3 $\frac{0.45}{0.09}$

4 0.8×0.07

5 $\frac{3}{0.5}$

6 $\frac{300}{500}$

7 $\frac{1}{0.4}$

8 0.06×200

9 0.7×5000

10 400×0.2

11 $\frac{12\,000}{5000}$

12 $\frac{200}{4000}$

13 400×0.006

14 $(0.3)^2$

15 500×300

16 80×6000

17 $(0.1)^2 \times 140$

18 $\frac{140}{(0.1)^2}$

Revision

Exercise 8 Mixed

1 0.6×0.5
2 1.5×0.3
3 $\frac{1.5}{0.3}$
4 150×3000
5 $\frac{150}{3000}$
6 $\frac{30}{1500}$
7 0.6×0.04
8 6000×40
9 $\frac{1}{0.25}$
10 $(0.4)^2$
11 $(0.4)^3$
12 $\frac{1}{250}$
13 0.05×1.2
14 500×1200
15 $\frac{1200}{500}$
16 $\frac{0.006}{0.0005}$
17 $\frac{12}{0.004}$
18 $\frac{1.5}{30}$

Part 3 Harder questions

Work the following.

Exercise 1

1 $\frac{3.2}{0.4}$
2 $\frac{4.8}{0.8}$
3 $\frac{6.0}{1.2}$
4 $\frac{0.35}{0.07}$
5 $\frac{1.6}{0.08}$
6 $\frac{0.6}{0.15}$
7 $\frac{0.8}{0.02}$
8 $\frac{1.4}{0.07}$
9 $\frac{2.0}{0.1}$
10 $\frac{1}{0.02}$
11 $\frac{3}{0.05}$
12 $\frac{0.001}{0.00025}$

Exercise 2

1 $\frac{1.3}{0.3}$
2 $\frac{7.3}{0.6}$
3 $\frac{0.53}{0.2}$
4 $\frac{0.07}{0.03}$
5 $\frac{3.0}{0.4}$
6 $\frac{1}{0.3}$
7 $\frac{5}{0.6}$
8 $\frac{0.63}{1.2}$
9 $\frac{8.01}{1.1}$
10 $\frac{0.006}{0.04}$
11 $\frac{10}{0.9}$
12 $\frac{1.3}{0.008}$

Exercise 3

1 $\frac{1.2+2.4}{0.9}$
2 $\frac{0.5+3.7}{0.6}$
3 $\frac{0.23+0.55}{0.3}$
4 $\frac{1.72+4.68}{0.09}$
5 $\frac{0.42-0.02}{0.07+0.05}$
6 $\frac{7.2+0.35}{0.32-0.12}$
7 $\frac{0.8+1.55}{0.324-0.32}$
8 $\frac{2.6-1.45}{1.8-1.75}$
9 $\frac{1-0.002}{1-0.2}$
10 $\frac{1.4+7.15}{0.3}$
11 $\frac{0.057-0.0115}{0.05} - 0.17$
12 $\frac{1}{1-0.95} - 0.95$

Revision

Exercise 4

1 $\frac{3.6 \times 1.5}{0.45}$

2 $\frac{4.2 \times 1.2}{0.14}$

3 $\frac{4.50}{2.5 \times 0.9}$

4 $\frac{5.5 \times 0.18}{0.33 \times 2.5}$

5 $\frac{0.63 \times 0.4}{0.7 \times 0.18}$

6 $\frac{2.8}{2.1 \times 0.4}$

7 $\frac{0.240}{0.06 \times 1.6}$

8 $\frac{12 \times 0.45}{1.8}$

9 $\frac{0.14 \times 9}{0.6}$

10 $\frac{0.36}{0.012 \times 5}$

11 $\frac{(2.4)^2}{0.36 \times 1.6}$

12 $\frac{0.6}{(0.02)^2 \times 9}$

13 $\frac{(0.056)^2}{(0.007)^2}$

14 $\frac{0.017 \times 0.06}{(0.1)^2 \times 0.34}$

15 $\frac{1}{(0.1)^2 \times (0.2)^2}$

Exercise 5

1 $\frac{1.2}{0.2}$

2 $\frac{3.2}{0.4}$

3 $\frac{4.8}{0.12}$

4 $\frac{3.95 + 0.25}{0.6}$

5 $\frac{2.78 - 1.23}{0.05}$

6 $\frac{0.465 - 0.15}{0.05}$

7 $\frac{15.3 + 3.4}{1.1}$

8 $\frac{37.5}{2.83 - 0.33}$

9 $\frac{847 - 208}{107.6 - 17.6}$

10 $\frac{2 - 0.08}{1 - 0.92}$

11 $\frac{2 + 1.2 + 0.05}{0.07 - 0.045}$

12 $\frac{2 \times (6.7 - 1.3)}{0.096 + 0.024}$

Exercise 6

1 $\frac{1.5 \times 3.6}{2.7}$

2 $\frac{8.1 \times 1.2}{3.6}$

3 $\frac{4.2 \times 2.1}{1.8 \times 1.4}$

4 $\frac{1.75 \times 1.5}{0.5 \times 0.42}$

5 $\frac{0.8 \times 0.27}{9 \times 0.012}$

6 $\frac{2.25}{0.5 \times 0.45}$

7 $\frac{7.2}{0.2 \times 1.8}$

8 $\frac{0.49 \times 5}{3.5 \times 0.14}$

9 $\frac{0.36}{1.8 \times 0.02}$

10 $\frac{0.036 \times 12.8}{0.9 \times 0.016}$

11 $\frac{2.4 \times 0.15 \times 4.2}{0.048 \times 1.8}$

12 $\frac{0.3 \times 12.5 \times 3.5}{7.5 \times 2.1 \times 0.5}$

Exercise 7

1 $\frac{9.61 + 2.89}{2 \times 0.25}$

2 $\frac{2.67 + 1.53}{0.14 \times 0.3}$

3 $\frac{3.5 \times 1.2}{1.75 + 0.35}$

4 $\frac{63.7 - 15.7}{0.5 \times 1.6}$

5 $\frac{0.173 - 0.023}{0.3 \times 0.4}$

6 $\frac{(0.48)^2}{6.4 \times 0.009}$

7 $\frac{13 - 0.2}{(1.6)^2}$

8 $\frac{1 - 0.005}{1 - 0.95}$

9 $\frac{5 + 0.2 - 0.04}{5 \times 0.2 \times 0.04}$

10 $\frac{5.4 \times (1 - 0.64)}{1.8 \times 0.03}$

11 $\left(\frac{0.41 + 0.23}{0.016}\right)^2$

12 $\left(\frac{1.3 + 0.14}{1.2}\right)^2$

Approximations

Rounding off

1 Measure the length of each of the following lines to the nearest mm.

a

b

c

d

e

f

g

h

2 Measure the lengths of the same lines to the nearest cm.

U V W X Y Z

3 (i) Measure the distances between the following pairs of points to the nearest mm.

a	UV	b	UW	c	UY	d	VW
e	VY	f	WZ	g	XZ	h	YZ

(ii) Measure the same distances to the nearest cm.

4 Find the perimeter of each of the following shapes and give your answers to the nearest cm.

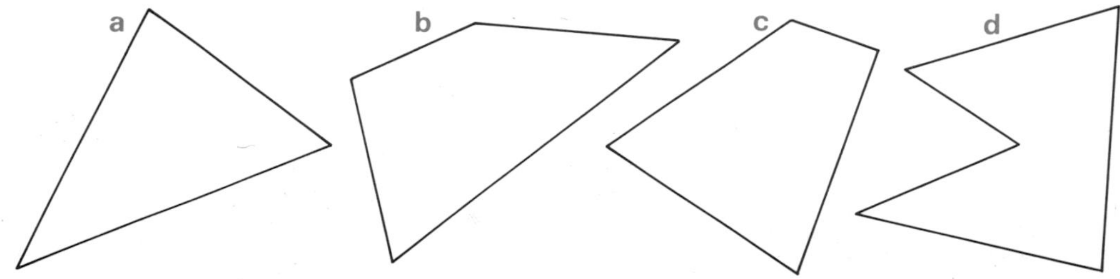

5 Write the following correct to the nearest cm. (Remember that 5 is rounded upwards.)

a	7.9 cm	b	3.1 cm	c	12.8 cm	d	23.2 cm
e	1.87 cm	f	7.18 cm	g	9.92 cm	h	48.5 cm
i	98.7 cm	j	189.6 cm	k	245.3 cm	l	60.2 cm

6 Write the following correct to the nearest £.

a	£3·98	b	£5·04	c	£8·77	d	£12·31
e	£36·42	f	£87·61	g	£14·52	h	£425·60
i	£392·50	j	£392·49	k	£0·78	l	£1·50

7 Write
a 12.6 cm correct to the nearest cm
b £64·82 correct to the nearest £
c 1.79 litres correct to the nearest litre
d 69.4 g correct to the nearest gram

e 14.7 kg correct to the nearest kg
f 371.3 cm^2 correct to the nearest cm^2
g 579 g correct to the nearest ten grams
h £271 correct to the nearest £10
i 437 cm^3 correct to the nearest 10 cm^3
j 3592 kg correct to the nearest 10 kg
k 798 mm correct to the nearest 100 mm
l 1803 correct to the nearest 100 cm
m 2623 litres correct to the nearest 100 litres
n 982 g correct to the nearest 100 grams
o 6.71 m correct to the nearest $\frac{1}{10}$ metre
p 2.38 litres correct to the nearest $\frac{1}{10}$ litre
q 13.75 kg correct to the nearest $\frac{1}{10}$ kg
r 0.852 km correct to the nearest $\frac{1}{10}$ km.

8 Write these numbers correct to 3 significant figures.
a 28.72　b 167.9　c 4.168　d 0.2037
e 0.06188　f 0.3427　g 10.85　h 6008.4
i 3155.1　j 819.7　k 14.96　l 0.1099

9 Write these numbers correct to 1 decimal place.
a 2.84　b 3.67　c 12.15　d 0.648
e 0.104　f 0.072

Write these numbers correct to 3 decimal places.
g 0.1234　h 0.9876　i 0.0025　j 0.4702
k 0.2065　l 0.0008

10 Copy and complete this table.

Number	correct to 2 significant figures	correct to 2 decimal places
6.829		
4.173		
3.025		
47.197		
82.566		
19.702		
0.627		
0.825		
0.101		
0.0876		
0.0209		
0.0655		

Approximations

11 Change these fractions to decimals and give your answers to the accuracy required.

a $\frac{1}{3}$ correct to 3 decimal places
b $\frac{2}{3}$ correct to 3 decimal places
c $\frac{4}{7}$ correct to 3 significant figures
d $\frac{1}{12}$ correct to 2 significant figures
e $\frac{1}{12}$ correct to 2 decimal places
f $\frac{1}{11}$ correct to 2 decimal places
g $\frac{1}{11}$ correct to 2 significant figures
h $\frac{5}{16}$ correct to 3 decimal places

12 One ball-bearing has a mass of 8.75 grams. Find the mass of 125 similar ball-bearings correct to

a the nearest gram
b the nearest 10 grams
c two significant figures.

13 One tin of beans has a mass of 0.376 kg. Find the mass of 48 tins correct to:

a the nearest kilogram
b one decimal place
c three significant figures.

14 Find the area of a rectangle 32.3 cm long and 13.6 cm wide. Give your answer correct to:

a the nearest cm^2
b the nearest 10 cm^2
c one decimal place
d one significant figure.

15 A coffee table is 1.65 m long and 0.45 m wide. Find its area correct to:

a the nearest $\frac{1}{10}$ m^2
b two decimal places
c three significant figures.

16 Find the area of this triangle, correct to:

a the nearest cm^2
b the nearest 10 cm^2
c two significant figures
d one decimal place.

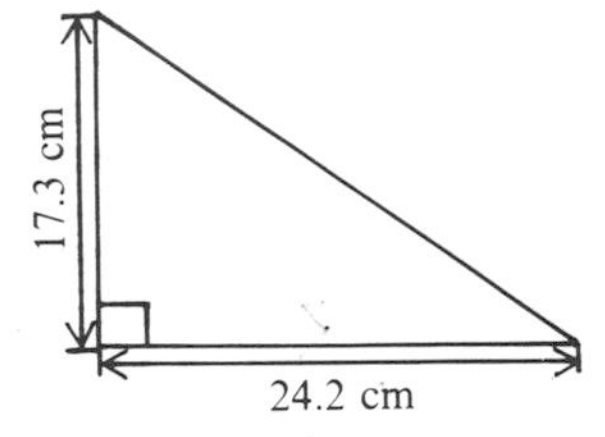

17 A triangle has a base of length 6.23 cm and a height of 5.74 cm. Calculate its area correct to:

a the nearest cm^2
b the nearest $\frac{1}{10}$ cm^2
c one significant figure
d one decimal place.

18 A car averages a speed of 48.7 miles per hour. How far does it travel in 12 hours. Give your answer correct to:

a the nearest 100 miles
b two significant figures.

19 An athlete averages a speed of 0.37 km per minute. What distance does he cover in 13.5 minutes? Give your answer correct to:

a the nearest kilometre
b one decimal place
c three significant figures.

20 7 cm^3 of metal has a mass of 51 grams. Find the mass of 1 cm^3 of this metal, correct to:

a the nearest gram
b one decimal place.

Limits of accuracy

21 What are the greatest and least possible prices of an object which costs

a £7 to the nearest pound
b £23 to the nearest pound
c £360 to the nearest ten pounds
d £420 to the nearest ten pounds
e 60p to the nearest ten pence
f £1·30 to the nearest ten pence
g £700 to the nearest hundred pounds
h £3400 to the nearest hundred pounds
i £10 to the nearest pound
j £10 to the nearest ten pence?

22 What are the greatest and least possible masses of an object which has a mass of

a 68 grams to the nearest gram
b 137 grams to the nearest gram
c 240 grams to the nearest gram
d 240 grams to the nearest ten grams
e 490 grams to the nearest ten grams
f 700 grams to the nearest hundred grams
g 3900 grams to the nearest hundred grams
h 2.7 grams to the nearest $\frac{1}{10}$th gram
i 4.0 grams to the nearest $\frac{1}{10}$th gram
j 4.00 grams to the nearest $\frac{1}{100}$th gram?

Addition

23

A B C D

The length AB is 12 cm to the nearest cm.
The length BC is 23 cm to the nearest cm.
The length CD is 8 cm to the nearest cm.

a Find the greatest possible length of ABCD.
b Find the least possible length of ABCD.

24 A man has three tins of paraffin holding 16 litres, 12 litres and 8 litres, all measured to the nearest litre. What is
a their greatest possible total volume
b their least possible total volume?

25 An empty biscuit tin has a mass of 640 grams. The lid of the tin has a mass of 80 grams. 780 grams of biscuits are placed in the tin. If all these masses are correct to the nearest ten grams. Find
a the greatest possible total mass
b the least possible total mass.

Approximations

26 The voltage drops across three resistors are 6.8 volts, 12.3 volts and 4.1 volts, all measured to the nearest $\frac{1}{10}$th volt. Find
- **a** the greatest possible value of the total of these voltage drops
- **b** the least possible value of this total voltage drop.

27 An equilateral triangle has one side measured as 6.8 cm to the nearest tenth of a centimetre. Find
- **a** the greatest possible value of its perimeter
- **b** the least possible value of its perimeter.

6.8 cm

28 An isosceles triangle has AB = AC = 26 mm and BC = 38 mm, both measurements correct to the nearest mm. Find
- **a** the greatest possible value of its perimeter
- **b** the least possible value of its perimeter.

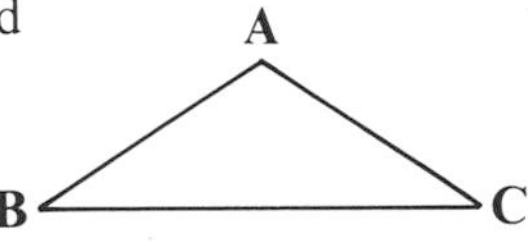

29 A rectangular field is 130 m long and 100 m wide, both measured to the nearest ten metres. Find
- **a** the greatest possible value of its perimeter
- **b** the least possible value of its perimeter.

30 Two long pieces of wood with a mass of 240 grams each and two shorter pieces with a mass of 200 grams each are used to make a rectangular picture frame. If both masses are to the nearest ten grams. Find
- **a** the greatest possible mass of the frame
- **b** its least possible mass.

Subtraction

31

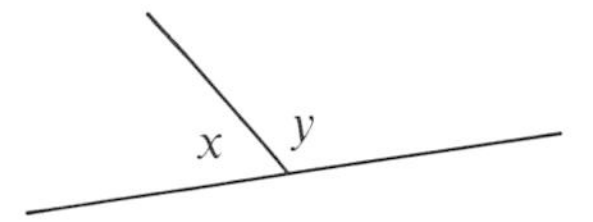

Angle x is measured to the nearest degree and found to be 52°. Find
- **a** the greatest possible value of angle y
- **b** the least possible value of angle y.

32 If angle m is 146° to the nearest degree. Find the greatest and least possible values of angle n.

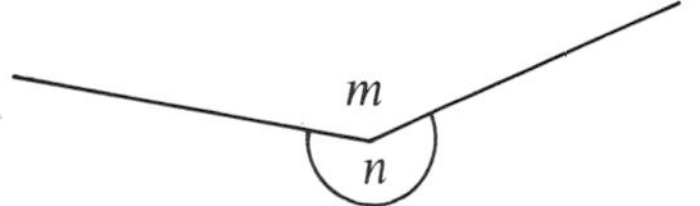

33

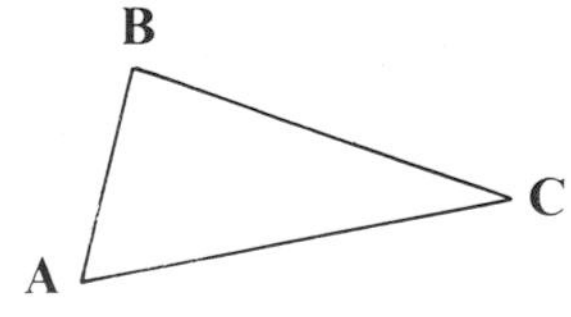

In △ ABC, $\widehat{A} = 63°$ and $\widehat{C} = 46°$, both measured to the nearest degree. Find
- **a** the greatest possible value of angle B
- **b** the least possible value of angle B.

34 A quadrilateral PQRS has $\widehat{P} = 84°$, $\widehat{Q} = 107°$ and $\widehat{R} = 134°$, all these angles being measured to the nearest degree. Find the range of possible values of $\widehat{S}$.

35 If AB and AC are measured to the nearest cm with the results shown. Find
- **a** the greatest possible length of BC
- **b** the least possible length of BC.

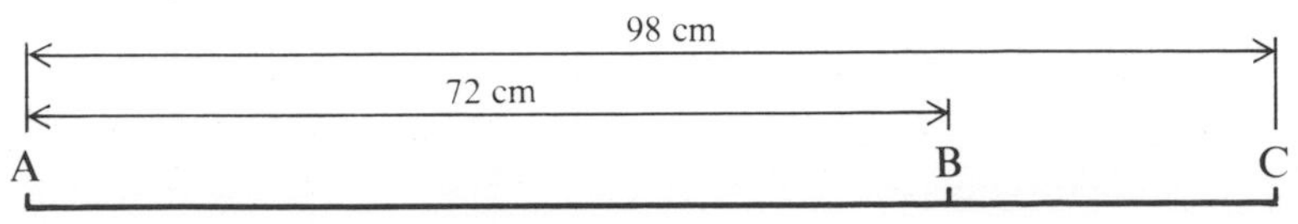

Approximations

36 If PR = 36.2 cm and QR = 12.4 cm (both measured to the nearest 0.1 cm), find the greatest and the least possible length of PQ.

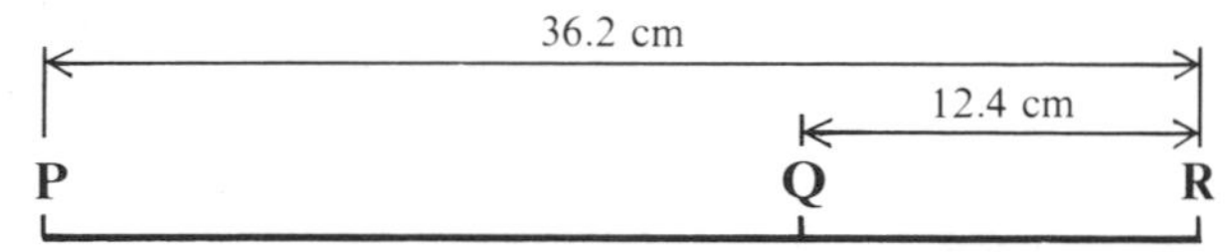

37 The temperature of a liquid is measured as 67.4°C and this temperature falls by 13.2°. If both these are measured to the nearest 0.1 of a degree, find
- a the greatest possible value of the final temperature
- b the least possible value of the final temperature.

38 A bag holding 18.6 kg of flour has 7.2 kg taken out, both these masses being accurate to the nearest $\frac{1}{10}$ of a kilogram. Find the range of possible values of the mass of flour left in the bag.

39 A beaker holds 367 cm^3 of water, and 124 cm^3 are poured out. If both these measurements are to the nearest whole cm^3, find the range of possible values of the volume of water left in the beaker.

40 A piece of string 8.45 metres long, has a 2.83 m length cut off. If both these measurements are correct to the nearest 0.01 m, find the range of possible values of the length which is left.

Multiplication

41 A rectangle is 17 cm long and 13 cm wide, both being measured correct to the nearest cm. Find
- a the largest possible value of its area
- b the smallest possible value of its area.

Give your answers correct to 2 significant figures.

42 A rectangular piece of hardboard is cut 46.1 cm long and 20.8 cm wide. If both these measurements are correct to 0.1 cm, find (to 3 significant figures) the largest and the smallest possible values of its area.

43 A boy measured one side of a square to the nearest millimetre, and his result was 124 mm. Find (to 3 significant figures) the largest possible value of the area of the square.

44 A meter measures the speed at which water flows into a tank. It reads 1.4 m^3 per minute, correct to the nearest 0.1 m^3. If the water flows for 35 minutes (correct to the nearest minute), what is (to 2 significant figures)
- a the greatest possible volume of water which has run into the tank
- b the least possible volume of water which has run into the tank?

45 A cyclist travels at a steady speed of 360 metres per minute (correct to the nearest 10 metres). He keeps up this speed for 18.6 minutes (correct to 0.1 min). Give both your answers to 2 significant figures.
- a What is the maximum possible distance he travelled?
- b What is the minimum possible distance he travelled?

46 1 cm^3 of iron has a mass of 7.87 grams (to the nearest 0.01 g). A bar of iron has a volume of 15.2 cm^3 (to the nearest 0.1 cm^3). Calculate the greatest possible mass of this bar of iron, giving your answer to three significant figures.

Approximations

Division

47 A rectangle has an area of 64 cm^2 (to the nearest cm^2) and a width of 5.1 cm (to the nearest mm). Find (to 2 significant figures)
 a its longest possible length
 b its shortest possible length.

48 A rectangular sheet of glass has a length of 32 cm (to the nearest cm) and an area of 624 cm^2 (to the nearest cm^2). Calculate the least possible width of the glass, giving your answer to 2 significant figures.

49 A car travels a distance of 74 km (to the nearest km) in a time of 1.2 hours (to the nearest 0.1 h). Calculate in km/h (to 2 significant figures)
 a its greatest possible average speed
 b its least possible average speed.

50 A man runs 64 m (to the nearest whole metre) in a time of 8.3 seconds (to the nearest 0.1 s). Calculate his greatest possible average speed to 2 significant figures.

51 To the nearest m^2, the area of a carpet is 26 m^2, and the salesman prices it at £132 (to the nearest £). Find the least possible average cost per m^2 of carpet to 2 significant figures.

52 To the nearest thousand litres, 165000 litres of water flows through a pipe in 42 hours (to the nearest hour). Find the least average rate of flow of water in litres per hour to 2 significant figures.

A mixture

53 If $v = u + at$ where u, a and t are measured such that $12 \leqslant u \leqslant 15$ and $7 \leqslant a \leqslant 9$ and $3 \leqslant t \leqslant 5$, find
 a the greatest possible value of v
 b the least possible value of v.

54 Given that $w = Ri^2$ and that $11 \leqslant R \leqslant 12$ and $2 \leqslant i \leqslant 3$, find the range of possible values of w.

55 Given that $A = \frac{1}{2}h(a + b)$ and that $2 \leqslant h \leqslant 6$ and $1.2 \leqslant a \leqslant 1.3$ and $2.7 \leqslant b \leqslant 2.8$, find the range of possible values of A.

56 Given that $R = \frac{v}{i}$ and that $8 \leqslant v \leqslant 12$ and $2 \leqslant i \leqslant 4$, find
 a the least possible value of R
 b the greatest possible value of R.

57 If $v = e - ri$ where $50 \leqslant e \leqslant 55$ and $8 \leqslant r \leqslant 10$ and $2 \leqslant i \leqslant 4$, find the maximum and minimum possible values of v.

58 If $p = qr - s$ where $6 \leqslant q \leqslant 9$ and $8 \leqslant r \leqslant 12$ and $21 \leqslant s \leqslant 25$, find the largest possible value of p.

59 Find the smallest possible value of s if $s = \frac{d}{t}$ where $15 \leqslant d \leqslant 18$ and $3 \leqslant t \leqslant 6$.

60 $x = \frac{y + z}{c}$ where $6 \leqslant y \leqslant 8$ and $10 \leqslant z \leqslant 12$ and $2 \leqslant c \leqslant 4$. Find the least possible value of x.

Approximations

61 Given that $p = \frac{m-n}{k}$, find the maximum possible value of p when $16 \leqslant m \leqslant 20$ and $8 \leqslant n \leqslant 10$ and $1 \leqslant k \leqslant 2$.

62 $s = \frac{v^2 - u^2}{2g}$ where $10 \leqslant v \leqslant 11$ and $7 \leqslant u \leqslant 8$ and $10 \leqslant g \leqslant 11$. Find the greatest possible value of s.

63 A beaker containing water has a mass of 156.7 g (to the nearest 0.1 g). After some time, their total mass is reduced to 154.9 g (to the same accuracy). What is the range of possible values of the mass of water lost by evaporation?

64 A circle has a radius of 6.2 cm (measured to the nearest 0.1 cm). Take $\pi = 3.14$ and use $A = \pi r^2$ to find the largest possible value of its area, to 2 significant figures.

65 A circle has a circumference of 23 cm (measured to the nearest cm). Take $\pi = 3.14$ and use $C = \pi d$ to find the smallest possible length of its diameter to 2 significant figures.

66 A square has an area of 72 cm^2 (to the nearest cm^2). Calculate the range of possible values of the length of its sides, giving your answer to 3 significant figures.

67 A cube has its volume measured as 47 cm^3 (to the nearest whole cm^3). Calculate the range of possible values of the length of its edges, giving your answer to 3 significant figures.

68 A rectangle has a perimeter of 84 cm and a length of 26 cm, both measured to the nearest cm.
Find the largest possible value of its width.

69 The perimeter and length of a rectangle are 22.8 cm and 8.3 cm respectively, where both are measured to the nearest millimetre.
Find the smallest possible width of the rectangle.

70 A simple pendulum of length l metres takes t seconds to make one complete swing.
Use the formula $g = \frac{4\pi^2 l}{t^2}$ to find the range of possible values of g when $\pi = 3.14$ and $l = 4.39$ m (to the nearest cm) and $t = 4.2$ seconds (to the nearest 0.1 second).
Give your answer correct to one decimal place.

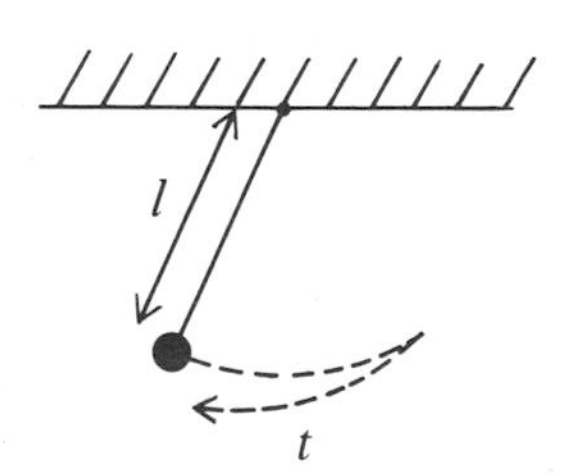

Decimals, Fractions and Percentages

Interchanging fractions and decimals
Percentages, decimals and fractions
Finding a percentage of a quantity
Cost price and selling price
Profit and percentage profit
Percentage error
Perimeter
Circumference of circles
Lengths of arcs

Interchanging fractions and decimals

Part 1 Comparing sizes

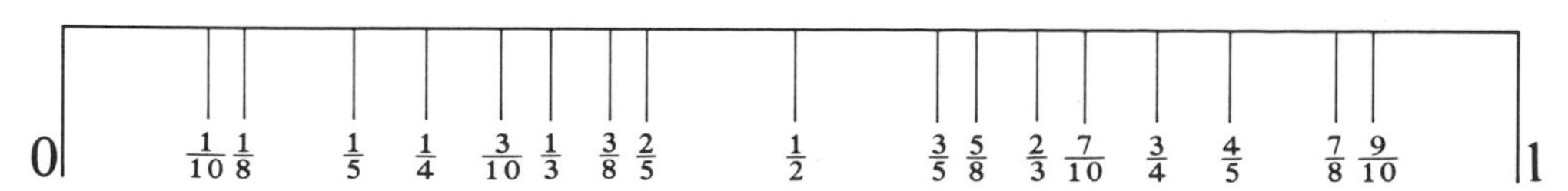

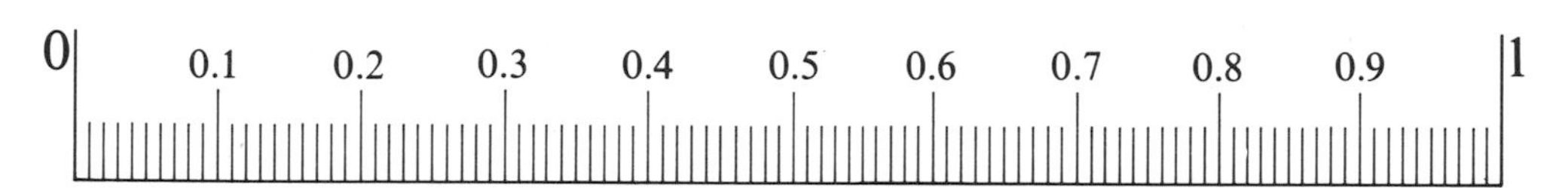

Use this diagram to answer the following questions.

1 Which is the bigger fraction in each of these pairs?

a $\frac{9}{10}$ and $\frac{7}{8}$ b $\frac{4}{5}$ and $\frac{3}{4}$ c $\frac{1}{2}$ and $\frac{3}{5}$

d $\frac{1}{3}$ and $\frac{3}{10}$ e $\frac{2}{3}$ and $\frac{5}{8}$ f $\frac{3}{8}$ and $\frac{2}{5}$

2 The sign $>$ means 'is bigger than'.
The sign $<$ means 'is smaller than'.

Copy these pairs of fractions and write the correct sign in the space between them.

a $\frac{3}{10}$ $\frac{1}{4}$ b $\frac{1}{2}$ $\frac{2}{5}$ c $\frac{2}{3}$ $\frac{3}{5}$

d $\frac{3}{10}$ $\frac{3}{8}$ e $\frac{7}{10}$ $\frac{4}{5}$ f $\frac{3}{8}$ $\frac{2}{3}$

g $\frac{3}{8}$ $\frac{1}{2}$ h $\frac{7}{8}$ $\frac{7}{10}$ i $\frac{3}{5}$ $\frac{3}{8}$

j $\frac{2}{3}$ $\frac{3}{4}$

3 Write the four fractions in ascending order of size (i.e. smallest first).

a $\frac{1}{8}$ $\frac{1}{10}$ $\frac{2}{5}$ $\frac{1}{4}$ b $\frac{1}{2}$ $\frac{3}{8}$ $\frac{7}{10}$ $\frac{2}{3}$

c $\frac{1}{2}$ $\frac{3}{10}$ $\frac{1}{5}$ $\frac{3}{8}$ d $\frac{3}{4}$ $\frac{3}{5}$ $\frac{7}{8}$ $\frac{1}{2}$

4 Which is the bigger decimal in each of these pairs?

a 0.2 and 0.4 b 0.2 and 0.04 c 0.84 and 0.87

d 0.05 and 0.5 e 0.3 and 0.29 f 0.07 and 0.2

g 0.6 and 0.61 h 0.1 and 0.09

5 Copy these pairs of decimals and write either $>$ or $<$ in the space between them.

a 0.8 0.4 b 0.07 0.7 c 0.19 0.2

d 0.9 0.09 e 0.7 0.6 f 0.07 0.6

g 0.07 0.06 h 0.1 0.09 i 0.1 0.99

j 0.89 0.98

Interchanging fractions and decimals

6 Write the four decimals in order of size, smallest first.

a	0.3	0.09	0.25	0.38
b	0.74	0.7	0.07	0.69
c	0.6	0.06	0.66	0.59
d	0.09	0.1	0.01	0.11

Line-up a ruler or set square across the diagram as shown here, then answer the questions which follow.

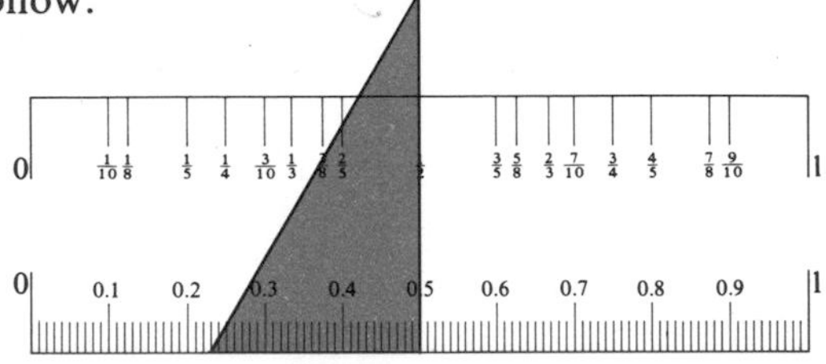

7 Write the fractions which are equal to these decimals.

a 0.1 b 0.2 c 0.3 d 0.4 e 0.5
f 0.6 g 0.7 h 0.8 i 0.25 j 0.75

8 Write the fractions which are *almost* equal to these decimals.

a 0.33 b 0.12 c 0.37 d 0.67 e 0.62
f 0.87

9 Write the decimals which are equal to these fractions.

a $\frac{1}{2}$ b $\frac{9}{10}$ c $\frac{3}{10}$ d $\frac{1}{10}$ e $\frac{2}{5}$
f $\frac{4}{5}$ g $\frac{1}{5}$ h $\frac{3}{5}$

10 Copy these decimals and fractions and write $>$, $<$ or $=$ in the space between them.

a $\frac{1}{4}$ 0.02 b $\frac{1}{2}$ 0.5 c 0.8 $\frac{4}{5}$
d 0.5 $\frac{1}{5}$ e $\frac{1}{3}$ 0.3 f $\frac{1}{4}$ 0.4
g 0.25 $\frac{1}{4}$ h 0.9 $\frac{9}{10}$ i $\frac{2}{3}$ 0.6
j $\frac{3}{4}$ 0.75 k 0.2 $\frac{1}{2}$ l 0.2 $\frac{1}{5}$

Part 2 Changing fractions to decimals

1 **Examples** $\frac{3}{10} = 0.3$ $\frac{39}{100} = 0.39$ $\frac{239}{1000} = 0.239$

Change the following fractions to decimals.

a $\frac{7}{10}$ b $\frac{9}{10}$ c $\frac{1}{10}$ d $\frac{49}{100}$ e $\frac{57}{100}$
f $\frac{91}{100}$ g $\frac{7}{100}$ h $\frac{9}{100}$ i $\frac{1}{100}$ j $\frac{7}{1000}$
k $\frac{9}{1000}$ l $\frac{1}{1000}$ m $\frac{243}{1000}$ n $\frac{777}{1000}$ o $\frac{101}{1000}$
p $\frac{999}{1000}$ q $\frac{99}{1000}$ r $\frac{21}{1000}$ s $\frac{3}{1000}$ t $\frac{1}{1\,000\,000}$

Interchanging fractions and decimals

2 Example $\frac{3}{8} = 8\overline{)3.0^60^40}$ (quotient 0.3 7 5) $= 0.375$

Change these fractions to terminating decimals.

a	$\frac{5}{8}$	b	$\frac{3}{4}$	c	$\frac{1}{4}$	d	$\frac{7}{8}$	e	$\frac{4}{5}$
f	$\frac{3}{5}$	g	$\frac{1}{8}$	h	$\frac{2}{5}$	i	$\frac{1}{5}$	j	$\frac{3}{20}$
k	$\frac{7}{20}$	l	$\frac{9}{20}$	m	$\frac{11}{25}$	n	$\frac{13}{25}$	o	$\frac{21}{25}$
p	$\frac{19}{20}$	q	$\frac{13}{40}$	r	$\frac{21}{40}$	s	$\frac{49}{50}$	t	$\frac{5}{16}$
u	$\frac{15}{16}$	v	$\frac{3}{25}$	w	$\frac{1}{20}$	x	$\frac{1}{50}$	y	$\frac{1}{16}$

3 Copy these fractions and change them to decimals by using your answers to question 2. **Learn them by heart.**

$\frac{1}{2}$

$\frac{1}{4}$ $\frac{3}{4}$

$\frac{1}{5}$ $\frac{2}{5}$ $\frac{3}{5}$ $\frac{4}{5}$

$\frac{1}{8}$ $\frac{3}{8}$ $\frac{5}{8}$ $\frac{7}{8}$

4 Example $\frac{1}{6} = 6\overline{)1.0^40^40^4}$ (quotient 0.1 6 6 ...) $= 0.1\dot{6}$

Change these fractions to recurring decimals.

a	$\frac{1}{3}$	b	$\frac{2}{3}$	c	$\frac{5}{6}$	d	$\frac{2}{9}$	e	$\frac{4}{9}$
f	$\frac{8}{9}$	g	$\frac{3}{11}$	h	$\frac{7}{9}$	i	$\frac{1}{9}$	j	$\frac{5}{11}$
k	$\frac{6}{11}$	l	$\frac{5}{9}$	m	$\frac{5}{12}$	n	$\frac{7}{12}$	o	$\frac{10}{11}$
p	$\frac{9}{11}$	q	$\frac{13}{30}$	r	$\frac{19}{30}$	s	$\frac{1}{7}$	t	$\frac{2}{7}$
u	$\frac{3}{7}$	v	$\frac{4}{7}$	w	$\frac{1}{12}$	x	$\frac{1}{11}$	y	$\frac{2}{11}$

5 Copy these fractions and change them to decimals by using your answers to question 4. **Learn them by heart.**

$\frac{1}{3}$ $\frac{2}{3}$

$\frac{1}{9}$ $\frac{2}{9}$ $\frac{4}{9}$

$\frac{5}{9}$ $\frac{7}{9}$ $\frac{8}{9}$

Part 3 Changing decimals to fractions

1 Examples $0.3 = \frac{3}{10}$ $0.43 = \frac{43}{100}$

Change these decimals to fractions.

a	0.7	b	0.9	c	0.1	d	0.47	e	0.99
f	0.13	g	0.57	h	0.03	i	0.01	j	0.07
k	0.371	l	0.999	m	0.449	n	0.001	o	0.003
p	0.017	q	0.049	r	0.103	s	0.809		

Interchanging fractions and decimals

2 Examples $0.8 = \frac{\cancel{8}^{4}}{\cancel{10}_{5}} = \frac{4}{5}$ $\quad 0.64 = \frac{\cancel{64}^{\cancel{32}\,16}}{\cancel{100}_{\cancel{50}\,25}} = \frac{16}{25}$

Change these decimals to fractions. Cancel as far as possible to give your answer in its lowest terms.

a	0.6	b	0.4	c	0.2	d	0.02	e	0.08
f	0.18	g	0.32	h	0.16	i	0.24	j	0.25
k	0.75	l	0.55	m	0.84	n	0.002	o	0.006
p	0.125	q	0.375	r	0.625	s	0.025	t	0.275
u	0.144	v	0.0008	w	0.0625	x	0.3125		

3 Example Change the recurring decimal $0.\dot{2}\dot{7}$ to a fraction.

Let $x = 0.272727\ldots$

then $100x = 27.2727\ldots$

Subtract $99x = 27.00000\ldots$

$x = \frac{27}{99} = \frac{3}{11}$

Change these recurring decimals to fractions.

a	0.666...	b	0.333...	c	0.777...	d	0.222...
e	0.181818...	f	0.121212...	g	0.363636...	h	0.636363...
i	0.010101...	j	0.1666...	k	0.8333...	l	0.08333...
m	0.41666...	n	0.0666...	o	0.1363636...	p	0.369369...
q	0.459459...						

Percentages, decimals and fractions

1 Find one word to fit the following descriptions. Each of your answers should begin with the letters C E N T.

- a An insect with many legs
- b One hundred years
- c A commander in the Roman Army in charge of 100 men
- d A scale of temperature based on 100 degrees
- e A person who lives a hundred years
- f A celebration of an event which happened 100 years ago
- g One hundred of these make one metre
- h One hundred of these make an American dollar

2 The number line from 0 to 1 can be labelled in three different but equivalent ways, using decimals, fractions and percentages.

0	0.1	0.2	0.3	0.4	0.5	0.6	0.7	0.8	0.9	1	**Decimals**
0	$\frac{1}{10}$	$\frac{2}{10}$	$\frac{3}{10}$	$\frac{4}{10}$	$\frac{5}{10}$	$\frac{6}{10}$	$\frac{7}{10}$	$\frac{8}{10}$	$\frac{9}{10}$	1	**Fractions**
0	10%	20%	30%	40%	50%	60%	70%	80%	90%	1	**Percentages**

Copy and complete this table. Use the diagram to help you.

Decimal	0.1	0.3	0.7						
Fraction				$\frac{9}{10}$	$\frac{4}{10}$	$\frac{8}{10}$			
Percentage							20%	50%	60%

Percentages to decimals

3 Change these percentages to decimals.

a	3%	b	5%	c	9%	d	12%	e	27%
f	36%	g	49%	h	87%	i	97%	j	106%
k	178%	l	205%	m	334%				

Decimals to percentages

4 Change these decimals to percentages.

a	0.02	b	0.07	c	0.08	d	0.15	e	0.23
f	0.39	g	0.41	h	0.92	i	1.02	j	2.09
k	4.13	l	0.7	m	0.5				

Percentages to fractions

5 Change these percentages to fractions.

a	3%	b	7%	c	29%	d	37%	e	41%
f	49%	g	51%	h	77%	i	99%	j	101%
k	103%	l	119%	m	201%				

Percentages, decimals and fractions

6 Change these percentages to fractions and cancel as far as possible.

a	20%	b	30%	c	40%	d	50%	e	60%
f	80%	g	25%	h	75%	i	15%	j	45%
k	35%	l	5%	m	95%	n	12%	o	16%
p	24%	q	48%	r	32%	s	36%	t	72%
u	150%	v	125%	w	250%	x	236%	y	155%
z	312%								

7 Change these percentages to fractions in their lowest terms.

a	$2\frac{1}{2}\%$	b	$7\frac{1}{2}\%$	c	$12\frac{1}{2}\%$	d	$37\frac{1}{2}\%$	e	$62\frac{1}{2}\%$
f	$87\frac{1}{2}\%$	g	$\frac{1}{2}\%$	h	$\frac{1}{3}\%$	i	$3\frac{1}{3}\%$	j	$13\frac{1}{3}\%$
k	$6\frac{2}{3}\%$	l	$33\frac{1}{3}\%$	m	$66\frac{2}{3}\%$	n	$3\frac{1}{5}\%$	o	$3\frac{1}{8}\%$
p	$9\frac{3}{8}\%$								

Fractions to percentages

Change these fractions to percentages.

8 First type

a	$\frac{11}{50}$	b	$\frac{9}{50}$	c	$\frac{17}{50}$	d	$\frac{31}{50}$	e	$\frac{7}{25}$
f	$\frac{12}{25}$	g	$\frac{16}{25}$	h	$\frac{21}{25}$	i	$\frac{11}{20}$	j	$\frac{3}{20}$
k	$\frac{13}{20}$	l	$\frac{19}{20}$	m	$\frac{7}{10}$	n	$\frac{9}{10}$	o	$\frac{1}{10}$
p	$\frac{3}{10}$	q	$\frac{2}{5}$	r	$\frac{4}{5}$	s	$\frac{3}{5}$	t	$\frac{1}{4}$
u	$\frac{3}{4}$	v	$\frac{1}{2}$	w	$\frac{104}{200}$	x	$\frac{62}{200}$	y	$\frac{48}{300}$
z	$\frac{168}{400}$								

9 Second type

a	$\frac{5}{8}$	b	$\frac{7}{8}$	c	$\frac{1}{8}$	d	$\frac{1}{6}$	e	$\frac{1}{3}$
f	$\frac{2}{3}$	g	$\frac{2}{9}$	h	$\frac{4}{9}$	i	$\frac{7}{9}$	j	$\frac{1}{7}$
k	$\frac{2}{7}$	l	$\frac{5}{7}$	m	$\frac{6}{7}$	n	$\frac{2}{11}$	o	$\frac{5}{11}$
p	$\frac{7}{11}$	q	$\frac{5}{12}$	r	$\frac{7}{12}$	s	$\frac{11}{12}$	t	$\frac{2}{15}$
u	$\frac{4}{15}$	v	$\frac{3}{16}$	w	$\frac{9}{16}$	x	$\frac{11}{16}$	y	$\frac{5}{21}$
z	$\frac{13}{24}$								

A mixture

10 Change to a decimal.

a 35% b 17% c 63% d 9% e 4% f 125%

11 Change to a percentage.

a 0.76 b 0.83 c 0.6 d 0.1 e 0.08 f 1.04

12 Change to a fraction.

a 9% b 19% c 89% d 3% e 23% f 143%

Percentages, decimals and fractions

13 Change to a fraction and cancel as far as possible.

a 90% b 55% c 25% d 28% e 48% f 175%

14 Change to a percentage.

a $\frac{21}{50}$ b $\frac{24}{25}$ c $\frac{1}{20}$ d $\frac{11}{100}$ e $\frac{11}{20}$ f $\frac{11}{25}$

15 Change to a percentage.

a $\frac{8}{9}$ b $\frac{3}{7}$ c $\frac{5}{6}$ d $\frac{3}{8}$ e $\frac{1}{12}$ f $\frac{7}{15}$

16 Copy and complete this table, cancelling all fractions as far as possible.

Decimal	0.25	0.2	0.35						
Fraction				$\frac{3}{4}$	$\frac{5}{9}$	$\frac{8}{11}$			
Percentage							85%	36%	120%

Finding a percentage of a quantity

Method 1 Using fractions

Example 18% of 125 metres $= \frac{\cancel{18}^{9}}{\cancel{100}_{\cancel{4}_{2}}} \times \cancel{125}^{5} = \frac{45}{2} = 22\frac{1}{2}$ metres

Work the following. Where necessary give your answer using a fraction.

1 25% of 140 metres **2** 35% of 60 metres **3** 12% of 75 tonnes

4 36% of 150 tonnes **5** 40% of 120 kg **6** 6% of 125 kg

7 14% of 75 metres **8** 18% of 225 metres **9** 15% of 35 litres

10 45% of 25 litres **11** 24% of 45 kg **12** 16% of 80 kg

13 60% of 18 kg **14** 48% of 20 kg **15** 32% of 55 metres

16 8% of a 35-metre roll of cloth is wasted. Find the length of the wasted material.

17 35% of the money raised in a raffle is given to a charity. If £140 is raised how much does the charity receive?

18 A bank account earns interest at 8%. What interest is earned if there is £225 in the account?

19 A box of chocolates has a mass of 250 grams. 15% of this mass is packaging. Find the mass of the packaging.

20 A petrol engine wastes 65% of the energy in the petrol. If the petrol contains 90 kJ of energy, how much is wasted?

21 A farmer grows potatoes on 15% of his land and silage on 24% of his land. If he has 300 hectares, on how many hectares does he grow
a potatoes **b** silage?

22 A boy spends 35% of his savings on sports equipment. If he had saved £70,
a how much is spent **b** how much is left?

23 43% of the population in a village are men and 37% are women. If the population is 2500, find the number of
a men **b** women **c** children.

24 A carpet covers 65% of a floor. If the area of the floor is 15 m^2, find
a the area covered by carpet **b** the area left uncovered.

25 45 m^2 of sheet metal is used to make an artefact but 16% of it is wasted. Find
a the area wasted **b** the area used.

26 14% of $12\frac{1}{2}$ litres **27** 15% of $17\frac{1}{2}$ kg **28** 32% of $7\frac{1}{2}$ tonnes

29 35% of $6\frac{2}{3}$ kg **30** 22% of $16\frac{2}{3}$ litres **31** 24% of $6\frac{1}{4}$ grams

32 32% of $11\frac{1}{4}$ metres **33** 64% of $9\frac{3}{8}$ km **34** 72% of $18\frac{1}{3}$ litres

35 45% of $28\frac{1}{3}$ kg **36** 36% of $43\frac{3}{4}$ tonnes **37** 105% of $9\frac{1}{6}$ tonnes

38 16% of a blend of tea comes from Sri Lanka, the remainder coming from India. How much tea in a $12\frac{1}{2}$ kg tea chest comes from
a Sri Lanka **b** India?

39 90% of $5\frac{1}{3}$ litres of fruit squash is water. How many litres of this squash are
a water **b** other ingredients?

40 70% of an alloy is copper. The rest is iron. If a piece of this alloy has a mass of $13\frac{1}{3}$ kg, what mass of it is
a copper **b** iron?

Finding a percentage of a quantity

41 A $37\frac{1}{2}$ litre cask is 84% full of wine.
- a How many litres of wine are in the cask?
- b How much more wine is needed to fill the cask?

42 A farmer treats 27% of his $53\frac{1}{3}$ hectares of land with insecticide. How many hectares are
- a treated
- b untreated?

43 A piece of elastic $31\frac{1}{4}$ cm long is stretched so that its length increases by 56%. Find
- a how far it is stretched
- b its new total length.

44 A car makes a journey of $112\frac{1}{2}$ km of which 44% is by motorway. How far is travelled
- a on motorways
- b on ordinary roads?

Method 2 Using decimals

Example Find 6% of £324

$$1\% = £\,3.24$$
$$6\% = £\,3.24 \times 6 = £19.44$$

1 Divide by 100 to work out the following.

a	1% of £642	b	1% of £589	c	1% of £794
d	1% of £3650	e	1% of £5900	f	1% of £84
g	1% of 824 metres	h	1% of 179 metres	i	1% of 47 metres
j	1% of 24.5 metres	k	1% of 48.2 kg	l	1% of 8.5 kg
m	1% of 7.0 litres	n	1% of 12 litres		

2 By finding 1% first, work out these percentages.

a	2% of £470	b	3% of £645	c	5% of £127
d	8% of £1350	e	7% of 5420 metres	f	4% of 862 metres
g	12% of 148 kg	h	14% of 64 kg	i	23% of 85 kg
j	32% of 23.4 kg	k	41% of 65.2 metres	l	53% of 124.5 metres
m	85% of 387.2 litres	n	73% of 2.5 litres	o	6% of 1.4 kg
p	24% of £3·50				

3 You invest £645 in a Building Society and after one year you get 8% interest. How much interest do you receive?

4 3% of the fuel used in a machine is wasted. If 237 litres of fuel are used, how much is wasted?

5 A farmer has 346 hectares of which 4% are not productive. How many hectares are not productive?

6 Four people have money in deposit accounts at a bank which gives 12% interest each year. For each person, find the interest they receive in one year.
- a Mrs Stubbs deposits £152.
- b Mr Slaithwaite deposits £724.
- c Miss Haynes deposits £1350.
- d Mr Higgs deposits £65.

7 A restaurant adds a service charge of 14% to its bills. If a party of people eat £48 of food, what service charge will they have to pay?

Finding a percentage of a quantity

8 Income tax is deducted from a man's wage at a rate of 32%. Find the tax he pays if
a £175 **b** £84 of his earnings is taxed.

9 A steel girder is 18 metres long. 12% of the girder is cut off. Find the length removed.

10 Mr Baldwin has £627 invested in a company which gives a bonus of 14% at the end of the year. How much is this bonus?

11 18% of the bolts made by a machine are faulty. If a total of 3250 bolts are made, how many are faulty?

12 3250 candidates sit a science examination. 22% of them fail and 6% of them get a distinction. How many
a fail the exam **b** get a distinction?

13 A man earns £625 a month and he is given a rise of 17%. How much more per month does he now earn?

14 An examination is marked out of 250. Alison scores 64% and Andrew scores 58%. Find the actual marks which each received.

15 35% of the length of the motorway between two access points is under repair. If the distance between the access points is 23.6 km, find the length of motorway being repaired.

16 Jim Spriggs earns £3·60 each week delivering newspapers. If he gets a rise of 15%, how much more does he get each week?

17 A market gardener has 6.5 hectares of land and he plants celery on 8% of his land and carrots on 14%. How many hectares are planted with
a celery **b** carrots?

18 In a recipe for a cake, 45% of the mass is flour and 15% of the mass is fruit. If all the ingredients have a mass of 2.6 kg, find the mass of
a the flour **b** the fruit.

19 26% of the area of a newspaper is given to advertising and 12% is used for photographs. If the area of the newspaper is 3.5 m^2, what is the area of
a the advertisements **b** the photographs?

20 Value Added Tax of 15% is charged on an article costing £4·80. How much tax is added?

21 Copy this layout and calculate $4\frac{1}{2}\%$ of £244.

$$\begin{aligned} 1\% &= £ \\ & \underline{\qquad\qquad \times 4} \\ 4\% &= £ \\ \tfrac{1}{2}\% &= £\underline{\qquad\qquad} + \\ 4\tfrac{1}{2}\% &= £\underline{\qquad\qquad} \end{aligned}$$

22 Use the same method to work these.

a $3\frac{1}{2}\%$ of £186 **b** $2\frac{1}{2}\%$ of £348 **c** $5\frac{1}{2}\%$ of £124
d $7\frac{1}{2}\%$ of £688 **e** $12\frac{1}{2}\%$ of £314 **f** $15\frac{1}{2}\%$ of £404
g $21\frac{1}{2}\%$ of £1462 **h** $37\frac{1}{2}\%$ of £2046 **i** $52\frac{1}{2}\%$ of £3280
j $87\frac{1}{2}\%$ of £460 **k** $47\frac{1}{2}\%$ of £72 **l** $28\frac{1}{2}$ of £46

Finding a percentage of a quantity

23 Calculate the value of the following.

a	$2\frac{1}{4}\%$ of £448	b	$5\frac{1}{4}\%$ of £852	c	$3\frac{1}{5}\%$ of £670
d	$6\frac{1}{5}\%$ of £1240	e	$4\frac{1}{3}\%$ of £1560	f	$8\frac{1}{3}\%$ of £2280
g	$12\frac{1}{4}\%$ of £172	h	$12\frac{3}{4}\%$ of £172	i	$23\frac{1}{3}\%$ of £381
j	$23\frac{2}{3}\%$ of £381	k	$35\frac{1}{4}\%$ of £64	l	$35\frac{3}{4}\%$ of £48

24 A Building Society offers investors $9\frac{1}{2}\%$ interest each year. Find how much interest the following people receive.
a Mr Balbus invests £648 for a year.
b Jimmy Hawkins keeps his savings of £58 for one year.
c Miss Parsons has £1250 invested for a year.

25 The same Building Society charges house buyers $14\frac{1}{4}\%$ interest on the mortgages they take out. Find the interest which these people have to pay.
a Mr and Mrs Guy borrow £9500.
b Miss Douglas borrows £6560.
c Mr Davey borrows £7640.

Using long multiplication and rounding-off

Example Find 5.1% of £36·25 to the nearest penny.

1% = £0·3625

5.1% = £0·3625
× 5.1
3625
18125
£1·84875 = £1·85 to the nearest penny

26 Work the following.

a	2.4% of £42·50	b	3.2% of £72·50	c	7.5% of £83·20
d	2.5% of £142·80	e	8.4% of £232·50	f	1.2% of £607·50
g	3.75% of £72·80	h	6.25% of £65·60	i	8.75% of £132·80
j	12.4% of £42·50	k	21.6% of £637·50	l	32.5% of £1468·40

27 Work the following correct to the nearest penny.

a	5.1% of £423·41	b	8.4% of £135·65	c	3.4% of £124·32
d	7.2% of £507·34	e	4.2% of £321·45	f	2.6% of £215·24
g	4.7% of £82·55	h	9.5% of £70·24	i	12.3% of £48·32
j	31.5% of £411·23	k	15.8% of £2104·30	l	24.1% of £207·45

28 A man is sent a bill for £45·36. If he settles it within a week, he will be given a 3.5% refund. How much is this refund worth?

29 In a sale, a store reduces the price of a table by 14.5%. If it cost £87·65 before the reduction, how much was it reduced by?

30 During one year the price of a new car increased by 12.3%. It cost £4256·50 at the start of the year.
a By how much has the price increased at the end of the year?
b What is the price at the end of the year?

Finding a percentage of a quantity

31 The cost of a package holiday was £285·45 last year. The same holiday this year is priced 8.75% higher. Find

a the increase in the price b the cost of the holiday this year.

32 You invest £1253·40 with a Building Society which gives you interest at a rate of 7.84% per annum.

a What interest do you earn in one year?

b What amount do you have altogether at the end of the first year?

33 A man spent £346·40 last year on train fares to and from work. The cost of rail travel increases by 23.5%.

a How much more will he have to pay if he travels the same amount this year?

b What will this year's train fares cost him?

34 A small firm export 42.5% of their products. They sold £65 430·00 of goods last year.

a How much was earned from exports?

b How much was sold on the home market?

(Give both answers correct to the nearest £.)

35 A large firm spent £86 570·00 on raw materials, 64.2% of which came from abroad. How much did they spend on raw materials which came

a from abroad b from home suppliers?

(Give both answers correct to the nearest £.)

Cost price and selling price

Selling price = Cost price + Profit or **Selling price = Cost price − Loss**

By cancelling

Find the profit or loss and the selling price when:

1 Cost price = £150 and percentage profit = 6%
2 Cost price = £350 and percentage profit = 8%
3 Cost price = £220 and percentage profit = 15%
4 Cost price = £140 and percentage profit = 5%
5 Cost price = £75 and percentage profit = 12%
6 Cost price = £125 and percentage profit = 16%
7 Cost price = £225 and percentage loss = 24%
8 Cost price = £450 and percentage loss = 3%
9 Cost price = £30 and percentage loss = 5%
10 Cost price = £110 and percentage loss = 15%.

11 A shop buys an old television for £150 and sells it to a customer making a profit of 11%. Find
a the profit b the selling price.

12 A furniture store makes a profit of 30% when a table is sold. If the table cost £45, find a the profit b the selling price.

13 A man buys a new car for £4200 and sells it a short time later making a loss of 12%. Find
a his loss b his selling price.

14 A firm makes garden gates for £70 and sells them to make a profit of 6%. Find
a its profit b its selling price.

15 A repair man buys an old vacuum cleaner for £32, mends it and sells it. If he makes a profit of 15%, for how much does he sell it?

16 A music shop bought records for £4 each but sells them making a loss of 30%. Find the selling price.

17 A sheet of glass is made for £55 and sold to make a profit of 5%. Find the selling price.

18 It costs a painter £45 to decorate the front of a house. If he adds on 35% profit, find the amount he charges.

By moving the decimal point

19 A store buys a cooker for £215 and sells it to make a profit of 8%.
a Find 1% of £215 by dividing by 100.
b Find 8% of £215.
c Find the selling price of the cooker.

Cost price and selling price

20 First find 1% of the cost price, then calculate the profit or loss and the selling price when:

a Cost price = £312 and percentage profit = 5%
b Cost price = £236 and percentage profit = 4%
c Cost price = £108 and percentage profit = 3%
d Cost price = £415 and percentage profit = 12%
e Cost price = £326 and percentage profit = 23%
f Cost price = £1240 and percentage profit = 31%
g Cost price = £2060 and percentage profit = 15%
h Cost price = £3140 and percentage loss = 35%
i Cost price = £550 and percentage loss = 14%
j Cost price = £175 and percentage loss = 6%.

21 Find the profit or loss and the selling price when:

a Cost price = £172·40 and percentage profit = 5%
b Cost price = £203·50 and percentage profit = 8%
c Cost price = £68·50 and percentage profit = 6%
d Cost price = £54·80 and percentage profit = 35%
e Cost price = £27·50 and percentage loss = 16%
f Cost price = £84·50 and percentage loss = 24%
g Cost price = £7·60 and percentage loss = 15%
h Cost price = £4·80 and percentage loss = 35%.

22 Find the profit or loss and the selling price, giving your answers correct to the nearest penny, when:

a Cost price = £214·30 and percentage profit = 4%
b Cost price = £331·70 and percentage profit = 3%
c Cost price = £405·10 and percentage profit = 9%
d Cost price = £113·70 and percentage profit = 17%
e Cost price = £64·25 and percentage profit = 21%
f Cost price = £40·56 and percentage profit = 34%
g Cost price = £51·68 and percentage loss = 6%
h Cost price = £176·40 and percentage loss = 12%
i Cost price = £418·80 and percentage loss = 24%
j Cost price = £8·73 and percentage loss = 15%.

23 A gas fire is bought for £72·50 and resold, making a profit of 14%. Find the profit and the selling price.

24 A shop buys a piano for £852·40 and sells it to make a 35% profit. Find the profit and the selling price.

25 A second-hand car is bought by a garage for £1244·50 and is resold at a profit of 8%. Find the selling price.

26 It cost Mrs Holdsworth £12·70 to knit two woollen jumpers for a friend. If she is to make a profit of 15%, how much should she sell them for?

Cost price and selling price

27 A crate of grapefruit costs a greengrocer £14·75 and he intends making a profit of 32%. How much will his customers pay for these grapefruit?

28 A camera shop buys colour film at £4·20 per roll. If it is sold at a profit of 15%, how much does each roll of film cost a customer?

29 A man buys sacks of potatoes from a farmer for £1·80 per sack. He sells them on the market making a profit of 35%. How much must he sell each sack for?

30 A shop has £275·50 worth of old stock. If it has a sale and reduces the prices by 24%, how much is this old stock sold for?

Answer the following correct to the nearest penny.

31 It costs Mr Jackson £32·35 to make a coffee table and he wants to make a profit of 26%. How much should he charge for it?

32 Garden fencing costs £12·85 per metre to manufacture. If a profit of 18% is to be made, how much will a buyer have to pay per metre?

33 An old-style bathroom suite costs a shop £628·30. The shop is prepared to make a 32% loss on it. What is the reduced price?

34 Paul buys a brand new bicycle for £114·72 and a few months later sells it, making a loss of 17%. How much does he sell it for?

35 It costs £32·70 for Mrs Woofer to bring up each of her pedigree pups. She sells them at a profit of 124%. What does she charge for her pups?

36 When Mr Lou Brickate bought shares in a North Sea oil firm it cost him £638·30, and then he sold them all at a profit of 132%. Find how much he sold them for.

Profit and percentage profit

Profit = Selling price − Cost price

$$\textbf{Percentage profit} = \frac{\textbf{Profit}}{\textbf{Cost price}} \times 100\%$$

(If the profit is negative, it is called a loss.)

For each of the following find **a** the profit **b** the percentage profit.

1 Cost price = £10 and selling price = £13
2 Cost price = £10 and selling price = £16
3 Cost price = £50 and selling price = £62
4 Cost price = £50 and selling price = £71
5 Cost price = £20 and selling price = £21
6 Cost price = £20 and selling price = £33
7 Cost price = £25 and selling price = £34
8 Cost price = £40 and selling price = £54
9 Cost price = £60 and selling price = £63

For each of the following find **a** the loss **b** the percentage loss.

10 Cost price = £25 and selling price = £22
11 Cost price = £30 and selling price = £27
12 Cost price = £80 and selling price = £76
13 Cost price = £120 and selling price = £105
14 Cost price = £80 and selling price = £61
15 Cost price = £160 and selling price = £154
16 Cost price = £240 and selling price = £213
17 Cost price = £48 and selling price = £39
18 Cost price = £125 and selling price = £122
19 Cost price = £625 and selling price = £585

20 A shop buys a transistor radio for £55 and sells it for £66. Find the percentage profit which the shop makes.
21 I buy a table for £28 and sell it for £35. Find my percentage profit.
22 A kitchen sink was bought for £36 and then sold again for £45. Find the percentage profit.
23 A large store buys an armchair for £144 and a customer is charged £225 for it. Find the percentage profit made by the store.
24 An electric cooker was bought for £240 and sold to a customer for £294. Find the percentage profit.
25 A tea-service cost £64 but the shop has a sale and sells it for £44. Find the percentage loss.
26 A mail-order firm sells typewriters costing £45 for only £30. Find the percentage reduction which it is offering.
27 A woman buys a second-hand sewing machine for £56. If a new one costs £72, find the percentage saving she has made.
28 A carpet costs £375 but a shop sells it for £455. Find the percentage profit.

Profit and percentage profit

29 A man buys a new car for £5250 and sells it after one year for £3850. Find his percentage loss.

With decimal division and cancelling

30 A tray of fresh peaches costs a trader £4·50 and he sells it for £5·40. Find his percentage profit.

31 An alarm clock is bought for £28·80 and sold for £36·00. Find the percentage profit.

32 A lady's coat is bought by a customer for £60·72 but it only cost the shop £52·80. Find the percentage profit.

33 A store has a sale in which a coffee table is sold for £37·62. If the table cost £39·60 originally, what percentage loss did the store make.

34 A shop buys a shirt for £9·60 and sells it for £11·28. Find the percentage profit.

35 If a greengrocer bought vegetables for £31·50 and sold them for £48·30, what is his percentage profit?

36 If a kettle cost the hardware shop £38·40 but in a sale is priced at £35·52, find the percentage reduction.

37 A profit of £4·20 is made on a garden gate which cost £12·60 to make. Find
a the selling price **b** the percentage profit.

38 A child's toy cost £5·04 to make and it is sold to make a profit of £3·36. Find
a the selling price **b** the percentage profit.

39 A spade leaves the factory for the shop at a price of £15·60, and the shop makes a profit of £1·95. Find
a the selling price in the shop **b** the percentage profit.

40 A manufacturer sells a radio to a shop for £35 though it only cost £25 to make. The shop then sells it to a customer for £36·40. Find the percentage profit made by
a the manufacturer **b** the shop.

41 A car tyre cost £22·50 to make and is sold for £30 to a garage, which then sells it to a car-owner for £35·70. Find the percentage profit made by
a the manufacturer **b** the garage.

Using logarithms or a calculator

42 You buy a record for £4·65 and the shop bought it for £3·47. Find
a the shop's profit **b** the percentage profit.

43 Mr Thorpe buys a new bicycle for £93·40 and he sells it two months later for £68·90. Find
a his loss **b** his percentage loss.

44 A shop pays £215 for an electric cooker and sells it to a customer for £287. Find the percentage profit the shop makes.

45 A man buys a used car for £476 but he then sells it, making a loss of £67. Find his percentage loss.

46 Mrs Mitchell buys some kitchenware for £87·30 on which the shop makes a profit of £14·90. Find
a the cost price **b** the percentage profit.

47 A garden centre sells a customer a lawn-mower for £37·80, making a profit of £9·50 on it. Find
a the cost price **b** the percentage profit.

Percentage error

$$\text{Percentage error} = \frac{\text{Actual Error}}{\text{Correct Value}} \times 100\%$$

1 A groundsman marks out the track for the 100-metre race, but the length he marks is 105 metres. Find
 a his error **b** his percentage error.

2 A matchbox has a label saying it holds 50 matches. The box you buy has 54 matches. Find
 a the error **b** the percentage error.

3 A tube of glue says it contains 50 cm³ of glue, but your tube holds 52 cm³. Find
 a the error **b** the percentage error.

4 A full bottle of ink holds 28 cm^3 of liquid. The label advertises that it contains only 25 cm^3. Find the error and the percentage error.

5 A plastic tape-measure stretches in the heat, and a 60-cm length of material is measured as only 57 cm. Find the error and the percentage error.

6 A speedometer on a car says the driver is travelling at 36 m.p.h. when in fact he is only moving at 30 m.p.h. Find the percentage error in the meter.

7 You travel the 150 mile journey to Nether Poppleton and your car's odometer has measured it as only 138 miles. Find its percentage error.

8 In a science experiment you are asked to use 300 cm^3 of water, but you in fact use 318 cm^3. What is your percentage error?

9 You think your electricity bill is going to be £63, but when it arrives you find it is only £56. Find your percentage error.

10 81° 27° 63°

A boy measures the three angles of a triangle using a protractor, with the results shown here.
 a What error has he made in the sum of the three angles?
 b What percentage error is this?

11 The label on a bottle of pills says that it holds 400 pills. When counted, there are 414 pills in the bottle. Find the percentage error.

12 Mrs Jones suspects her kitchen scales are inaccurate. A 360-gram bag of mixed fruit has a mass of 405 grams on the scales. Find the percentage error.

13 The chemistry teacher gives Tim Hale 720 cm^3 of liquid. When Tim measures it he says he has only got 693 cm^3. Find the percentage error in Tim's measurement.

14 A corridor is 270 cm wide. Carpet is bought for the floor but is found to be 18 cm too narrow. What is the percentage error in the width of the carpet?

15 At the top of Rachel's science paper is written 77%, but when she adds up the marks she finds she should have 84%. Find
 a the error in the teacher's marking
 b the percentage error.

Using decimals

16 A piece of string is known to be 6.40 cm long. A boy measures it to be 6.64 cm long. Find his percentage error.

17 Mrs Johnson's vegetable bill came to £3·60. She thought it should have been £3·45. Find her percentage error.

18 A room needs 31.5 m^2 of carpet. When Mr Smith worked out how much was needed he got 33.6 m^2. What was his percentage error?

19 Jennie adds up her savings to be £9·90. In fact she actually has £13·50. Find the percentage error in her arithmetic.

20 You estimate your gas bill to be £25·60. However, on arrival it is £28·80. What is your percentage error?

21 A workman writes down the readings from two dials.
Dial A 6.6 Dial B 15.8

Both these readings are inaccurate. Dial A should be 6.4 and dial B should be 16.0

a Find the workman's error for each dial.
b Find his percentage error for each dial.
c Which dial did he read more accurately?

22 A schoolboy checks the volumes of liquid in two bottles. The label on bottle A says 1.60 litres but in fact there are 1.55 litres in the bottle. The label on bottle B says 3.75 litres but in fact there are 3.80 litres in the bottle.
a Find the error in the labels on both bottles.
b Find the percentage error in the labels on both bottles.
c Which bottle is more accurately labelled?

Using logarithms or a calculator

23 Mrs Lamb asked for 9.50 kg of potatoes, but she got short-measure and was given only 8.24 kg. Find
a how much short she was
b the percentage error in the weight.

24 Mr Grimshaw has a mass of 86.5 kg but his bathroom scales give his mass as 73.2 kg. Find the percentage error in the scales.

25 An accurate thermometer measures room temperature as 21.5°C. A faulty thermometer measures the same temperature as 28.9°C. Find the percentage error in the faulty thermometer.

26 A broken window has an area of 1460 cm^2. Glass is ordered but is cut to the wrong size. If the sheet has an area of 1230 cm^2, find the percentage error made in cutting the glass.

27 A scientist compares an inaccurate voltmeter with a standardised voltmeter. The inaccurate one gave a reading of 42.6 volts while the accurate one said 44.1 volts. Find the percentage error in the inaccurate reading.

28 Lincoln to Inverness is 406 miles and Lincoln to Exeter is 241 miles. I travel both these journeys in different cars which measure the distances as 418 miles and 253 miles.
a Find the actual errors in the two odometers.
b Find the percentage errors in the two odometers.
c Which car has the more accurate odometer?

Percentage error

Longer problems

29 Angela uses her ruler and measures WX and XY. Her results are WX = 7.9 cm and XY = 5.2 cm. The correct lengths are WX = 8.0 cm and XY = 5.0 cm. Find

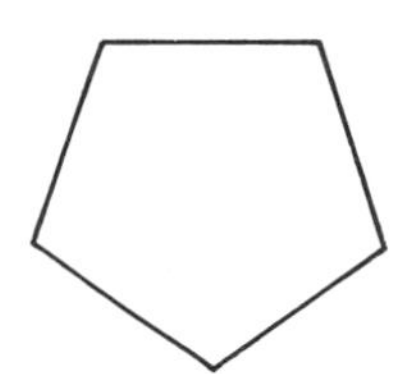

- a the correct area of the rectangle
- b the area of the rectangle using Angela's results
- c the error in her area
- d the percentage error (to 2 significant figures).

30 A boy uses a ruler to measure AB and BC and finds that AB = 4.2 cm and BC = 5.3 cm. In fact the actual lengths are exactly 4 cm and 5 cm. Find

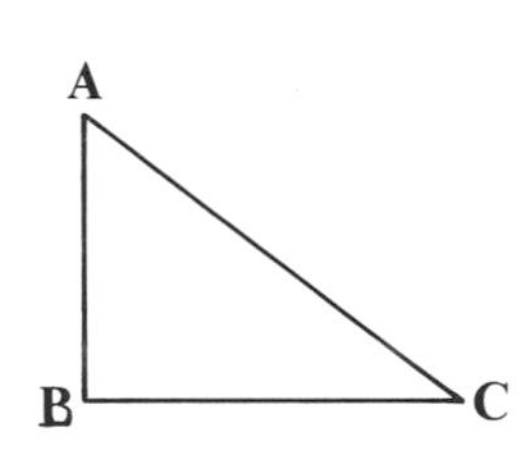

- a the correct area of triangle ABC
- b the area of triangle ABC using the boy's measurements
- c the error in this area
- d his percentage error (to 3 significant figures).

31 A regular pentagon has sides 2.4 cm long. Mark measures one side and says it is 2.6 cm long. Find

- a the actual perimeter of the pentagon
- b the perimeter using Mark's result
- c the percentage error in Mark's perimeter (to 2 significant figures).

32 (Take π as 3.14.)
A circle has a radius of 5 cm. You measure the radius inaccurately as 5.3 cm.

- a Calculate
  - (i) its actual circumference
  - (ii) the circumference using your inaccurate result
  - (iii) the error in your value of the circumference
  - (iv) the percentage error in this value.
- b Calculate
  - (i) its actual area
  - (ii) the area using your inaccurate result
  - (iii) the error in your value of the area
  - (iv) the percentage error in this area.

Give answers to 2 decimal places.

33 A rectangular box is known to be 24 cm long, 14 cm high and 15 cm wide. Barbara measures these lengths and gets 24.1 cm, 14.2 cm and 15.3 cm. Find

- a the correct volume of the box
- b the volume using Barbara's measurements
- c the error in her volume
- d the percentage error in her volume (to 2 significant figures).

34 A rectangular block 20 cm long has a square cross-section 12 cm by 12 cm. A boy measures the block and says the length is 20.1 cm long and the square has edges of 11.7 cm. Calculate

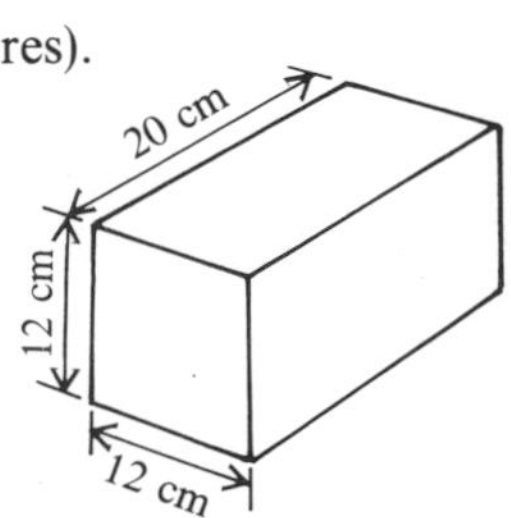

- a the true volume of the block
- b the volume using the inaccurate measurements
- c the error in this volume
- d the percentage error in this volume (to 3 significant figures).

Percentage error

35 A regular hexagon of side 7 cm fits inside a circle as shown. The perimeter of the hexagon can be taken as an approximate value for the circumference of the circle.

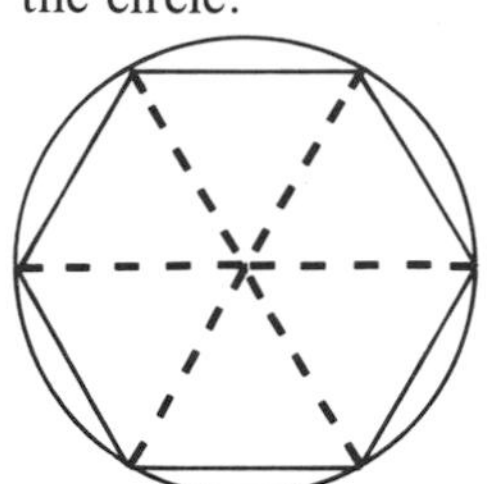

a Write down the radius of the circle.
b Write down the perimeter of the hexagon.
c Use $C = 2\pi r$ to find the circumference of the circle, taking $\pi = \frac{22}{7}$.
d What error is there in taking the hexagon's perimeter as the value of the circle's circumference?
e What percentage error is this (to 3 significant figures)?

36 A cylinder has a radius of 6.0 cm and a height of 14.0 cm. A schoolboy measures the radius of 5.8 cm and the height as 14.7 cm.
Take $\pi = \frac{22}{7}$ and calculate
a the true volume of the cylinder
b the volume using the boy's results
c the error in his volume
d the percentage error in his volume (to 2 significant figures).

37 The teacher writes $(x + y)^2$ on the blackboard. One pupil miscopies this as $(x - y)^2$.
a If $x = 9$ and $y = 1$ find (i) the value of $(x + y)^2$
(ii) the value which the pupil calculates
(iii) the pupil's percentage error.
b If $x = 4$ and $y = 1$ find (i) the value of $(x + y)^2$
(ii) the value which the pupil calculates
(iii) the pupil's percentage error.

38 A boy miscopies $\dfrac{p + q}{r}$ and writes $\dfrac{p - q}{r}$ instead.
a If $p = 34$, $q = 2$ and $r = 4$ find (i) the value of $\dfrac{p + q}{r}$
(ii) the value which the boy calculates
(iii) his percentage error.
b If $p = 48$, $q = 8$ and $r = 2$ find (i) the value of $\dfrac{p + q}{r}$
(ii) the value which the boy calculates
(iii) his percentage error.

39 A girl is trying to work out the value of $\sqrt{76.2}$ but she miscopies it and writes $\sqrt{72.6}$ instead. Find
a the value of $\sqrt{76.2}$
b the value she works out
c her percentage error.

40

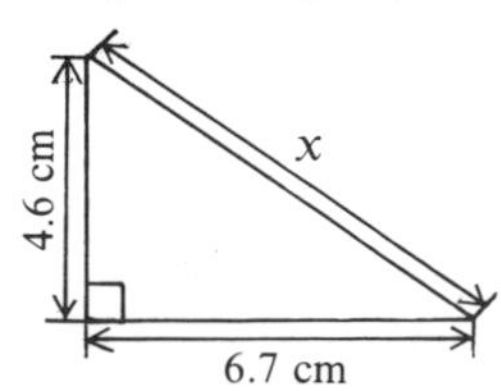

Richard tries to work out the length x in this right-angled triangle. Unfortunately he miscopies the lengths as 4.7 cm and 6.4 cm. Find
a the true length of x
b the length Richard calculates
c his percentage error.

Perimeter

1 Find the distance round the outside of these shapes.

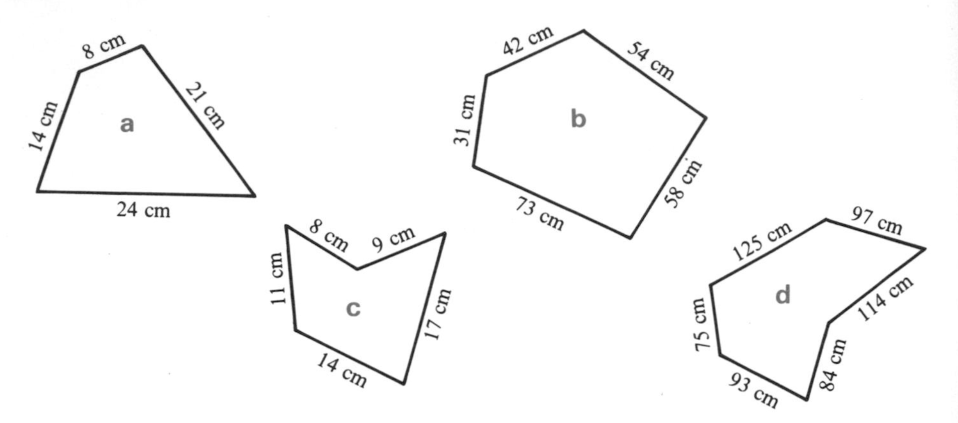

2 Calculate the perimeters of the following shapes.

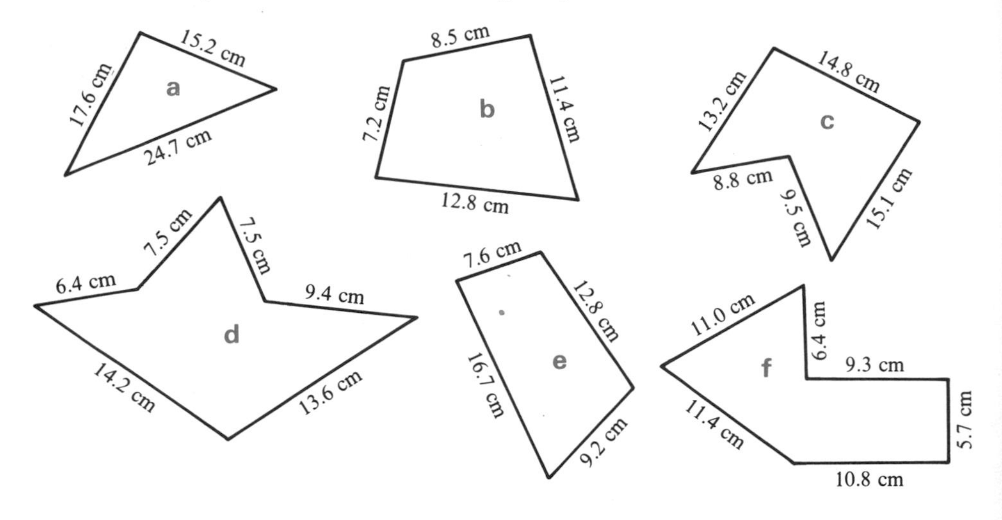

3 Use a ruler to measure the length of each side of these shapes to the nearest 0.1 cm. Then, calculate the perimeter of each shape.

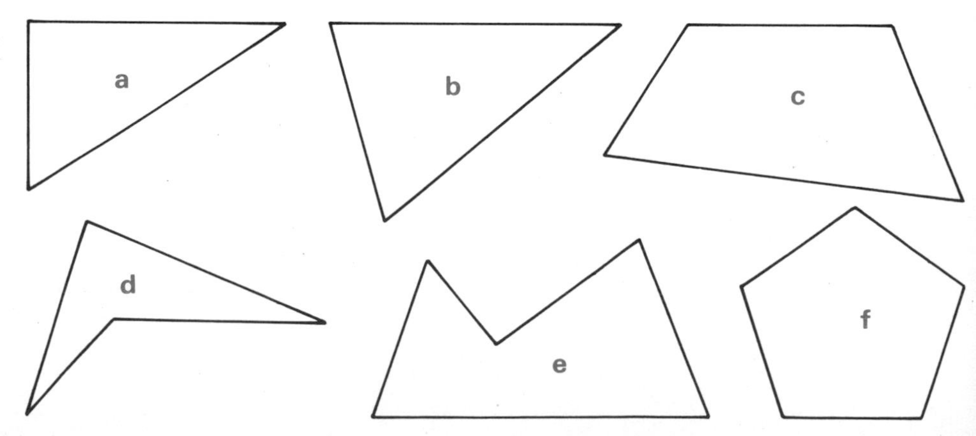

Perimeter

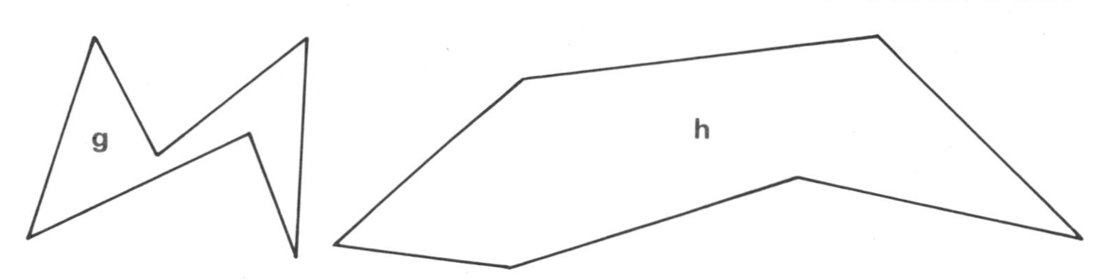

4 On cm^2 paper draw and label the x- and y-axes from 0 to 10.
Plot each set of points and join them together (in order) to make a shape.
Find the perimeter of each shape.

- **a** (1, 1) (4, 1) (4, 3) (3, 3) (3, 2) (1, 2) (1, 1)
- **b** (6, 1) (7, 1) (7, 3) (6, 3) (6, 4) (5, 4) (5, 1) (6, 1)
- **c** (8, 1) (10, 1) (10, 5) (9, 5) (9, 4) (8, 4) (8, 3) (9, 3) (9, 2) (8, 2) (8, 1)
- **d** (1, 3) (2, 3) (2, 4) (3, 4) (3, 5) (2, 5) (2, 6) (1, 6) (1, 5) (0, 5) (0, 4) (1, 4) (1, 3)
- **e** (0, 8) (2, 8) (2, 9) (5, 9) (5, 10) (1, 10) (1, 9) (0, 9) (0, 8)
- **f** (4, 5) (5, 5) (5, 6) (6, 6) (6, 8) (5, 8) (5, 7) (4, 7) (4, 8) (3, 8) (3, 6) (4, 6) (4, 5)
- **g** (7, 6) (10, 6) (10, 7) (9, 7) (9, 8) (10, 8) (10, 9) (7, 9) (7, 8) (8, 8) (8, 7) (7, 7) (7, 6)

5 On cm^2 paper draw and label both axes from 0 to 10.
Plot each set of points and join them together (in order) to make a shape.
Use a ruler to measure each side in centimetres to the nearest 0.1 cm.
Calculate the perimeter of each shape.

- **a** (1, 2) (4, 1) (3, 5) (1, 2)
- **b** (7, 1) (10, 2) (9, 4) (5, 3) (7, 1)
- **c** (2, 6) (4, 8) (3, 10) (0, 9) (2, 6)
- **d** (5, 5) (10, 6) (7, 7) (6, 8) (5, 5)
- **e** (5, 9) (9, 8) (10, 9) (8, 10) (5, 9)

6 Calculate the perimeters of these rectangles.

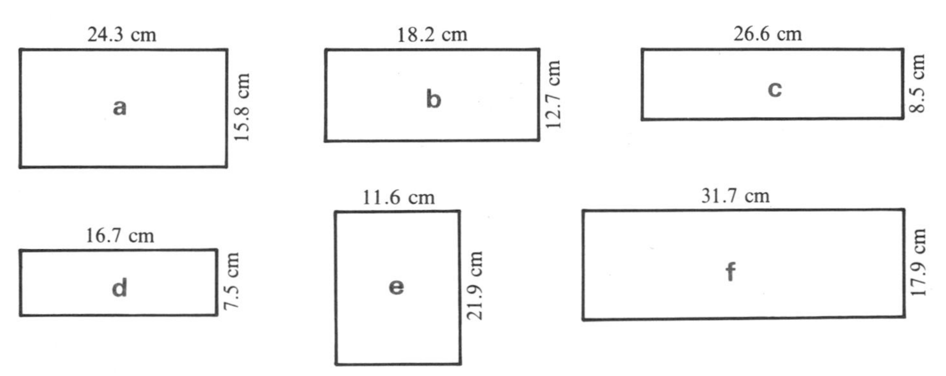

7 Calculate the perimeters of the ten rectangles given in this table.

	a	b	c	d	e	f	g	h	i	j
Length in cm	3.2	8.5	7.9	23.6	37.9	2.75	3.24	0.79	0.28	0.97
Width in cm	1.2	6.3	5.7	13.8	18.4	1.65	0.83	0.25	0.16	0.55

Perimeter

8 Use a ruler to measure the sides of these rectangles (to the nearest 0.1 cm) and so calculate their perimeters.

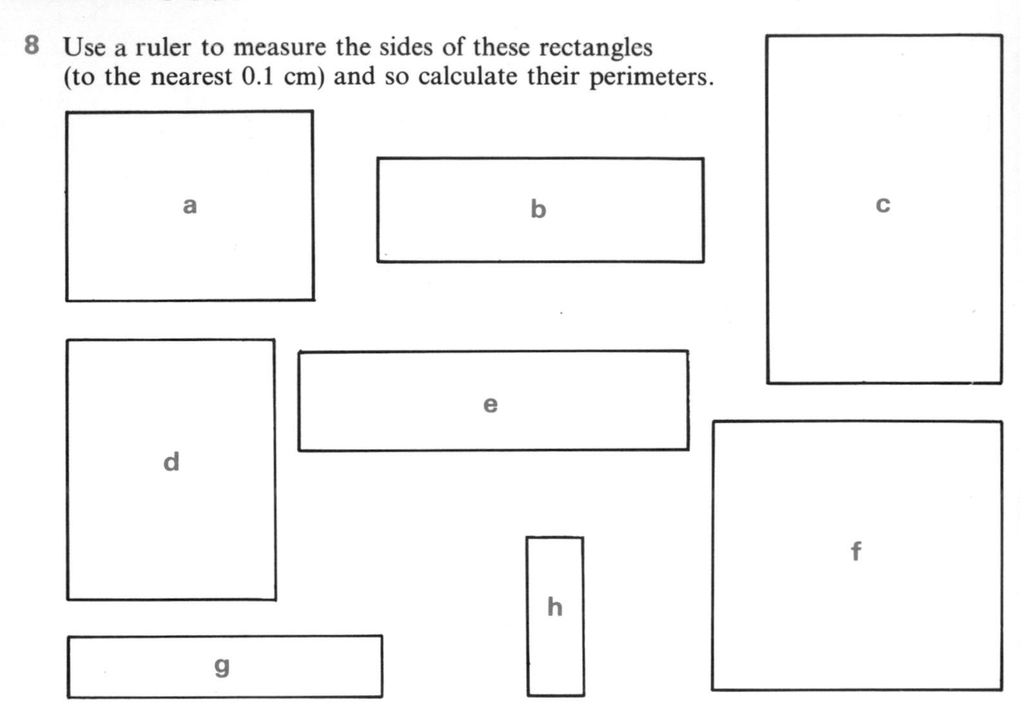

9 Use a piece of cotton to find the perimeters (in centimetres) of these shapes.

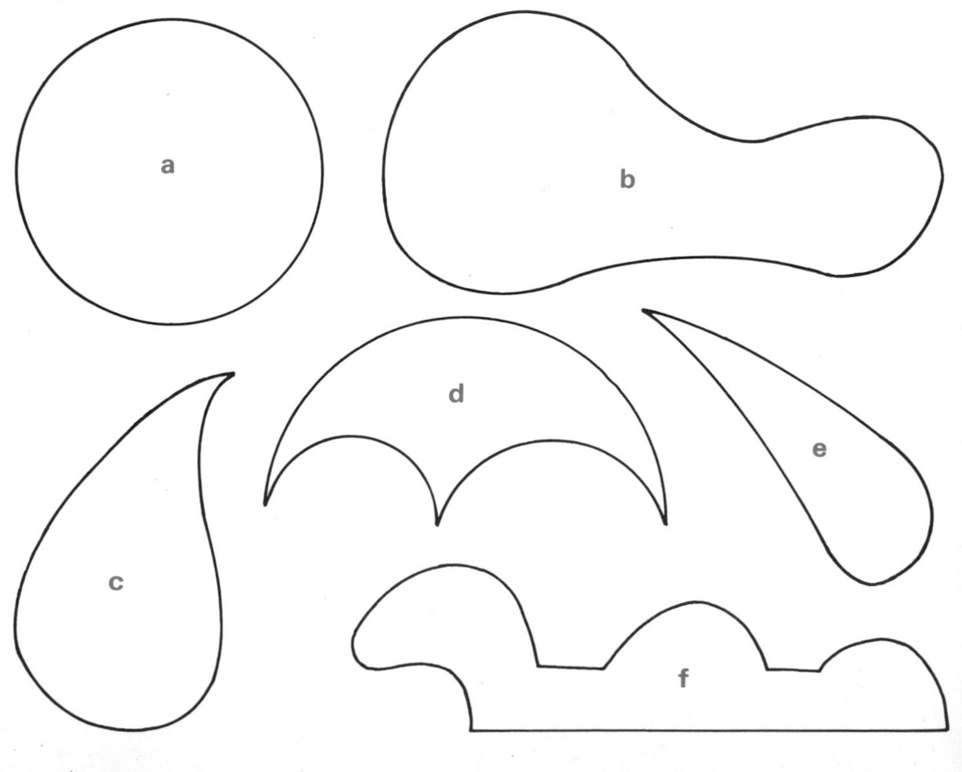

Perimeter

10 Calculate the perimeters of these shapes.

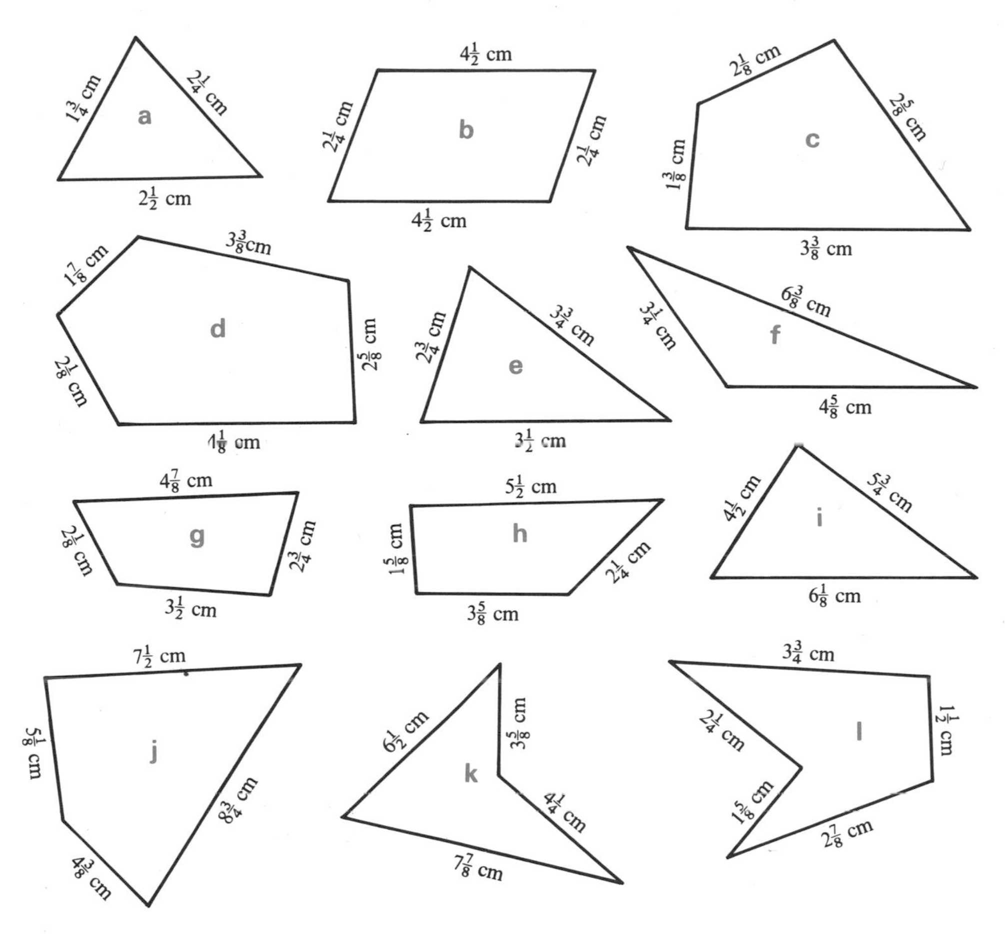

11 Calculate the perimeter of a triangle with sides of

- a $4\frac{1}{2}$ cm, $3\frac{3}{4}$ cm, $2\frac{1}{4}$ cm
- b $6\frac{1}{4}$ cm, $4\frac{1}{2}$ cm, $3\frac{1}{4}$ cm
- c $5\frac{1}{8}$ cm, $3\frac{1}{2}$ cm, $2\frac{1}{4}$ cm
- d $2\frac{3}{8}$ cm, $1\frac{1}{4}$ cm, $1\frac{1}{2}$ cm
- e $4\frac{3}{4}$ cm, $2\frac{1}{8}$ cm, $2\frac{3}{8}$ cm
- f $2\frac{2}{3}$ cm, $3\frac{1}{6}$ cm, $2\frac{5}{6}$ cm
- g $8\frac{1}{3}$ cm, $4\frac{1}{6}$ cm, $3\frac{1}{3}$ cm
- h $7\frac{5}{6}$ cm, $2\frac{1}{3}$ cm, $1\frac{5}{6}$ cm
- i $3\frac{1}{2}$ cm, $2\frac{1}{3}$ cm, $4\frac{1}{6}$ cm
- j $5\frac{2}{3}$ cm, $3\frac{5}{6}$ cm, $1\frac{1}{2}$ cm
- k $2\frac{5}{12}$ cm, $3\frac{2}{3}$ cm, $4\frac{1}{4}$ cm
- l $6\frac{3}{4}$ cm, $4\frac{7}{12}$ cm, $3\frac{1}{3}$ cm.

12 Calculate the perimeter of a rectangle with the following length and width

- a $2\frac{1}{2}$ cm and $1\frac{1}{4}$ cm
- b $4\frac{3}{4}$ cm and $3\frac{1}{2}$ cm
- c $5\frac{1}{8}$ cm and $2\frac{1}{4}$ cm
- d $1\frac{3}{8}$ cm and $1\frac{1}{4}$ cm
- e $3\frac{5}{8}$ cm and $2\frac{1}{2}$ cm
- f $2\frac{1}{3}$ cm and $1\frac{1}{6}$ cm
- g $4\frac{2}{3}$ cm and $3\frac{1}{6}$ cm
- h $5\frac{1}{3}$ cm and $2\frac{5}{6}$ cm
- i $3\frac{1}{12}$ cm and $2\frac{1}{4}$ cm
- j $7\frac{5}{12}$ cm and $4\frac{3}{4}$ cm
- k $6\frac{2}{3}$ cm and $3\frac{7}{12}$ cm
- l $5\frac{1}{3}$ cm and $1\frac{3}{4}$ cm

Perimeter

13 How long is the perimeter of
- a an equilateral triangle with sides of 8.4 cm
- b a square with sides of 6.3 cm
- c a regular pentagon with sides of 9.7 cm
- d a regular hexagon with sides of 12.5 cm
- e a regular octagon with sides of 15.4 cm
- f a regular decagon with sides of 6.75 cm?

14 a A square has a perimeter of 34.4 cm. How long is one of its sides?
- b A regular pentagon has a perimeter of 42.5 cm. How long is one of its sides?
- c A regular octagon has a perimeter of 108 cm. How long is one of its sides?

15 A young girl edges her poncho with a fringe costing £2·85 per metre length. Find
- a the perimeter of the poncho
- b the cost of the fringe she uses.

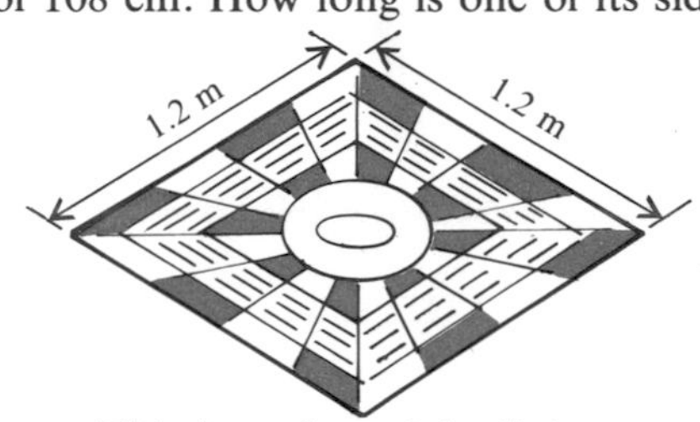

16

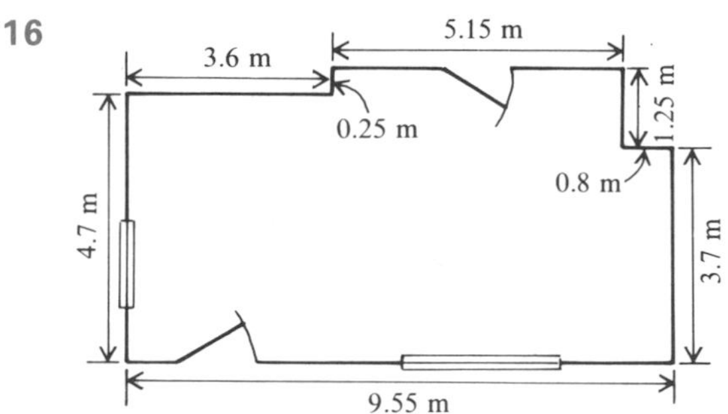

This is a plan of the living room of a house.
- a Calculate the perimeter of the room.
- b A carpet is fitted to cover all the floor. It is fixed round the edge of the room by strips of spikes costing 32 pence per metre length. Find the cost of the strips required.

17 A children's playground is edged with kerbstones which are each half a metre long and which cost £1·35 each. Calculate
- a the perimeter of the playground
- b the number of kerbstones required
- c the cost of the kerbstones required.

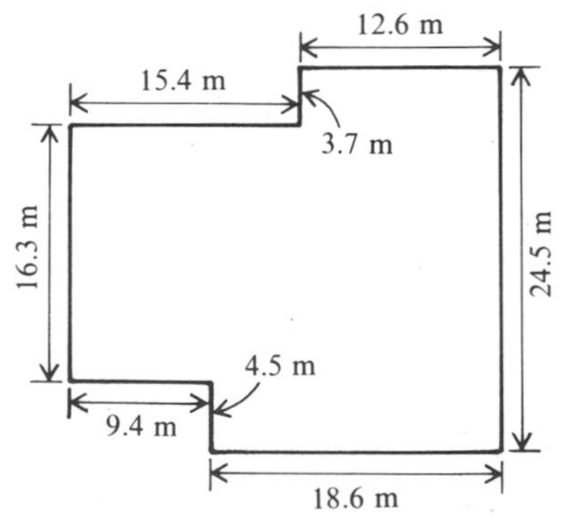

18

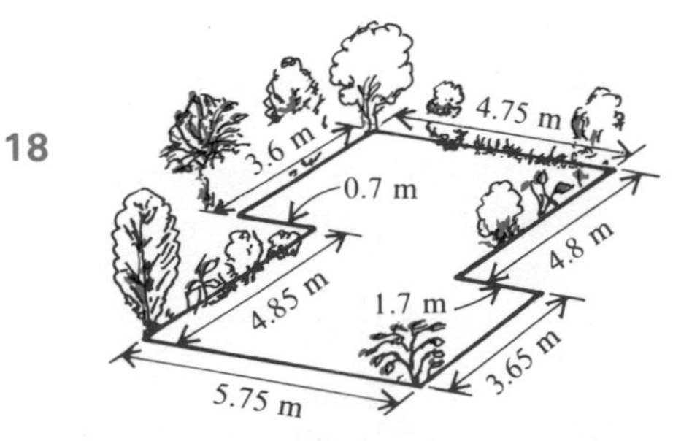

A gardener wants to edge his lawn with plastic hoops which are sold in 2-metre lengths costing 84 pence per length. Calculate
- a the perimeter of the lawn
- b the number of 2-metre lengths which he needs
- c the cost of the lengths needed.

Perimeter

19 This is a plan of one of the bedrooms in a new house and the builder fits it with a skirting-board around the bottom of its walls, leaving a gap of 0.9 m for the door. Calculate

a the perimeter of the floor of the room

b the length of skirting-board required

c the total cost of the skirting-board, if a 1-metre length costs £1·24.

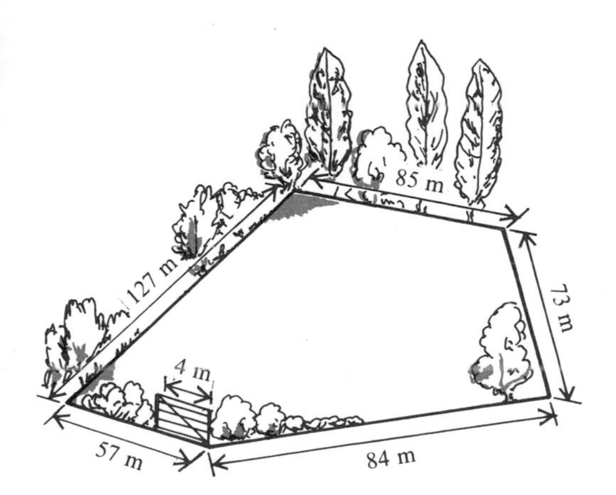

20 A farmer decides to fence around one of his fields, leaving a 4-metre gap for the gate.
He will need two stakes for each metre length of fencing. Each stake costs 65 pence and metre lengths of fencing cost £1·45. Calculate

a the perimeter of the field including the gate

b the length of fencing required and its cost

c the number of stakes required and their cost

d the total cost of stakes and fencing.

21 A car-park is being built.
Around its edge a rain-water channel is laid using $\frac{1}{2}$-metre lengths each costing 84 pence and with a mass of 6.5 kg. Find

a the perimeter of the car-park

b the number of $\frac{1}{2}$-metre lengths of channel which are needed

c the cost of the required channel

d the mass of this channel.

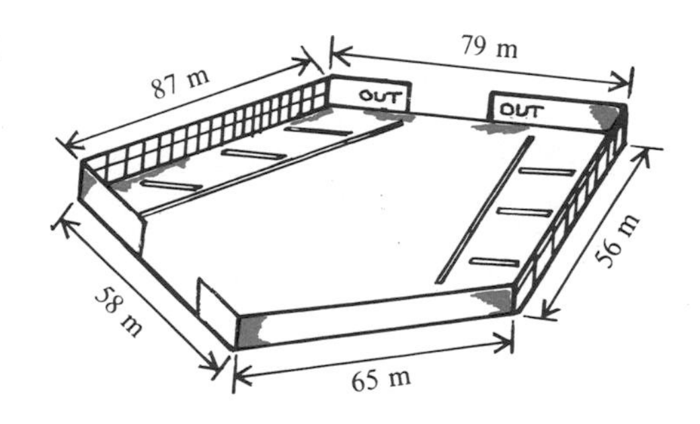

22 Jane makes a tablecloth 1.25 metres long and 0.75 metres wide. She hems around its edge, using a sewing machine which sews 4 stitches every centimetre. Calculate

a the perimeter of the tablecloth (i) in metres (ii) in centimetres

b the number of stitches in the hem around the edge

c the length of cotton thread in these stitches if the machine uses 1.5 cm of thread for each centimetre of hem

d the number of tablecloths which can be hemmed from one bobbin of cotton holding 120 metres of thread.

Circumference of circles

Introduction

Draw a circle of diameter 8 cm in your exercise book.

Measure its perimeter or circumference by stepping round it carefully with a piece of cotton.

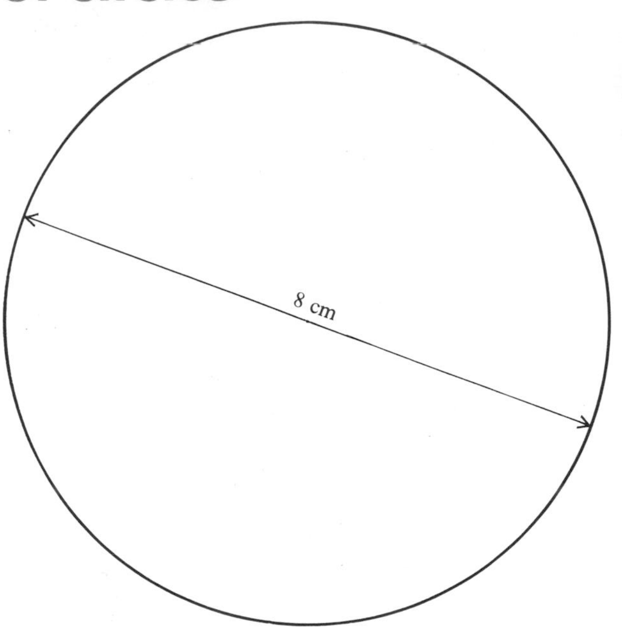

Copy the table below, enter your result and work out the value for the last column.

Repeat these instructions for circles with the diameters given in the table.

Diameter in cm	Circumference in cm	How many times longer is the circumference than the diameter?
8		
10		
12		
7		
6		
5		
4		
3		

Work out the average value of the numbers in the last column.

We see that the circumference is '3 and a bit' times longer than the diameter,

or, $C = 3 \text{ and a bit} \times d$
$= 3.14 \times d$
$= \pi d$
$= 2\pi r$ as $d = 2r$.

The value 3.14 is not exact. Another approximation is $\frac{22}{7}$. The exact value is given by the Greek letter π.

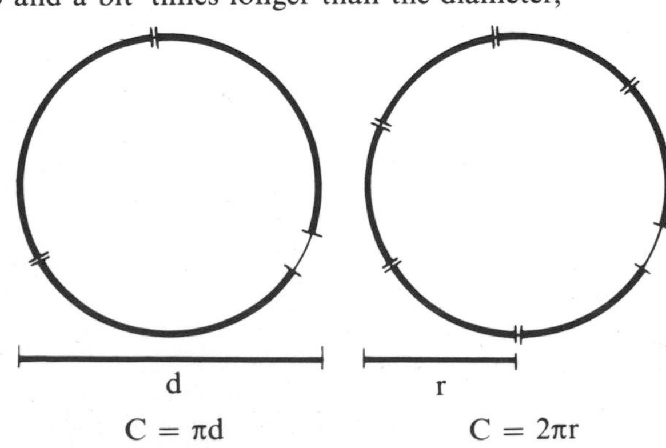

Circumference of circles

1 Use $C = \pi d$ to find the circumferences of the circles with these diameters. Take π as 3.14 and give your answers to 3 significant figures.

a	5 cm	b	4 cm	c	6 cm	d	3 cm
e	8 cm	f	12 cm	g	15 cm	h	18 cm
i	21 cm	j	23 cm	k	4.5 cm	l	8.2 cm
m	6.4 cm	n	7.1 cm	o	13.5 cm	p	21.6 cm

2 Use $C = 2\pi r$ to find the circumferences of the circles with these radii. Take π as 3.14 and give your answers to 3 significant figures.

a	2 cm	b	3 cm	c	5 cm	d	4 cm
e	6 cm	f	8 cm	g	12 cm	h	16 cm
i	$4\frac{1}{2}$ cm	j	$5\frac{1}{2}$ cm	k	$10\frac{1}{2}$ cm	l	3.6 cm
m	4.2 cm	n	7.8 cm	o	10.2 cm	p	12.6 cm

3 Use $C = \pi d$ to find the circumferences of the circles with these diameters. Take π as $\frac{22}{7}$.

a	14 cm	b	21 cm	c	28 cm	d	42 cm
e	70 cm	f	49 cm	g	35 cm	h	63 cm
i	$3\frac{1}{2}$ cm	j	$10\frac{1}{2}$ cm	k	$17\frac{1}{2}$ cm	l	$31\frac{1}{2}$ cm
m	$2\frac{1}{3}$ cm	n	$9\frac{1}{3}$ cm	o	$11\frac{2}{3}$ cm	p	$5\frac{1}{4}$ cm
q	$8\frac{3}{4}$ cm	r	$12\frac{1}{4}$ cm				

4 Use $C = 2\pi r$ to find the circumferences of the circles with these radii. Take π as $\frac{22}{7}$.

a	7 cm	b	14 cm	c	21 cm	d	35 cm
e	70 cm	f	1 cm	g	10 cm	h	20 cm
i	50 cm	j	$3\frac{1}{2}$ cm	k	$10\frac{1}{2}$ cm	l	$17\frac{1}{2}$ cm
m	$24\frac{1}{2}$ cm	n	$31\frac{1}{2}$ cm	o	2.1 cm	p	4.9 cm
q	8.4 cm	r	12.6 cm				

5 Use logarithms or a calculator to calculate the circumference of a circle with a diameter of

a	1.24 cm	b	2.05 cm	c	2.89 cm	d	5.65 cm
e	9.19 cm	f	24.6 cm	g	44.5 cm	h	80.6 cm
i	92.8 cm	j	165 cm	k	487 cm	l	675 cm.

6 Use logarithms or a calculator to calculate the circumference of a circle with a radius of

a	1.15 cm	b	1.33 cm	c	4.55 cm	d	8.15 cm
e	9.92 cm	f	15.8 cm	g	28.2 cm	h	48.5 cm
i	88.1 cm	j	132 cm	k	208 cm	l	555 cm.

Circumference of circles

In the following problems, take π as 3.14 unless otherwise instructed. Give answers to 3 significant figures where necessary.

7 A bell-tent has a diameter of 5.2 metres. Calculate the distance around the walls of the tent.

8 A windmill has sails each 8.6 metres long. Calculate how far the tip of a sail travels in one rotation.

9 The crankshaft of a bicycle is 17 cm long. If it turns once, what distance does the pedal at its end travel around the axle?

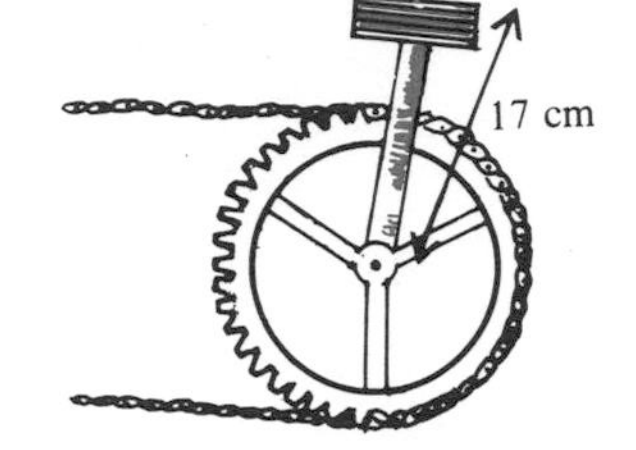

10 A clock has an hour-hand 2.5 cm long and a minute-hand 4.5 cm long. Calculate how far
- **a** the tip of the hour hand travels in 12 hours
- **b** the tip of the minute hand travels in 1 hour.

11 A circular mirror of diameter 35 cm is edged with gilt metal strip costing 4 pence per centimetre. Take π as $\frac{22}{7}$ to find
- **a** the length of strip needed
- **b** its cost.

12 A ship is tied to a quay by a rope wound round a capstan of radius 24 cm. If the rope is wound round a full six times, calculate
- **a** the circumference of the capstan
- **b** the length of the rope wound round it.

13 A roundabout in a major road has a radius of 14 metres. The edge of it is laid with kerbstones 50 cm long. Take π as $\frac{22}{7}$ to find
- **a** the circumference of the roundabout
- **b** the number of kerbstones needed.

14 The Earth has a radius of 6370 km. Calculate
- **a** the distance round the Equator
- **b** the distance from the North Pole to the South Pole.

15 A garden contains a circular pond with a diameter of $3\frac{1}{2}$ metres. Take π as $\frac{22}{7}$ and calculate
- **a** the circumference of the pond in metres and in centimetres
- **b** If tiles (20 cm long) are laid around the edge of the pond, how many will be needed?

16 A race-track has two straight sections 101 metres long joined by two semicircles of diameter 63 metres. Take π as $\frac{22}{7}$ and calculate
- **a** the distance around one semicircular end
- **b** the distance around the track
- **c** the number of laps which will be run in a 2-km race.

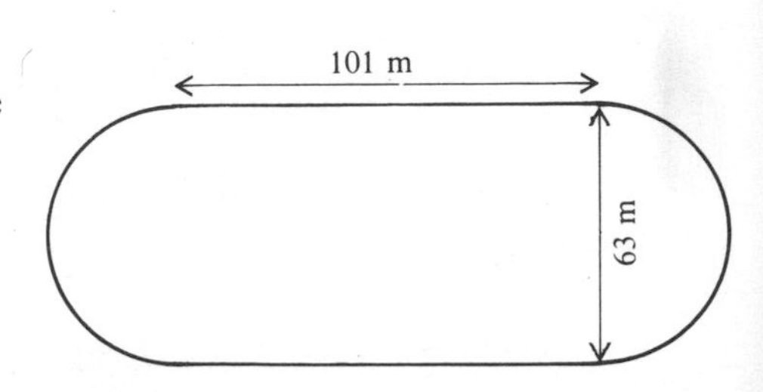

Circumference of circles

17 A race-track has the dimensions 145 m and 35 m as shown. Take π as $\frac{22}{7}$ and calculate

a the length of one lap

b the number of laps needed to complete a 3-km race.

18 A length of wire is wound 25 times around a drum of radius 30 cm. Calculate

a the circumference of the drum

b the length of wire in cm and in metres

c the cost of the wire (to the nearest £), if each metre length costs 45 pence.

19 A bucket is lowered into a well on a rope which winds around an axle of diameter 21 cm. Take π as $\frac{22}{7}$ and calculate

a the circumference of the axle

b the depth of the well, if the rope unwinds 35 times.

20 A 10p piece has a diameter of $2\frac{4}{5}$ cm. It is rolled without slipping along a flat surface. Take π as $\frac{22}{7}$ to find how far it travels forward in

a one full turn **b** 10 full turns.

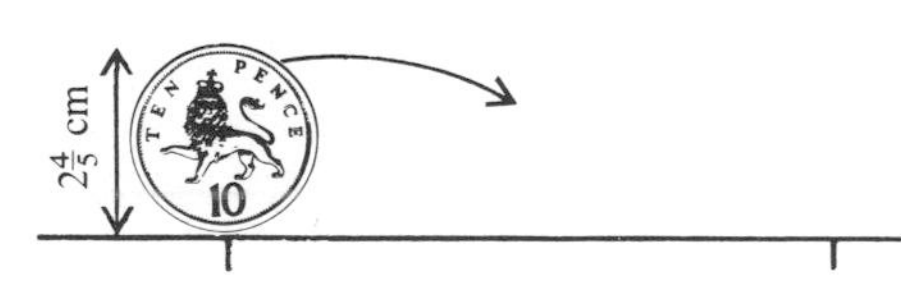

21 A bicycle has wheels of radius 30 cm. How far does the bicycle move forward if its wheels go round

a once **b** 25 times?

22 A bicycle has wheels of radius 20 cm. How far does it travel forward if its wheels go round

a once **b** 75 times?

23 A labourer on a building site pushes a large concrete drainage pipe of diameter 1.25 metres along the ground. How far does he push it if the pipe makes 12 full revolutions?

24 A length of cotton is wrapped round a reel 100 times. If the reel has a radius of 1.5 cm, find

a its circumference **b** the length of the cotton.

25 A bobbin of cotton has a radius of $1\frac{3}{4}$ cm. Take π as $\frac{22}{7}$ and calculate

a the circumference of the bobbin,

b how many times a piece of cotton $5\frac{1}{2}$ metres long can be wrapped around it.

26 A bobbin has a radius of $3\frac{1}{2}$ cm. 11 metres of thin string is wound around it. Take π as $\frac{22}{7}$ to find

a the circumference of the bobbin

b the number of times the string can be wound around it.

Circumference of circles

27 If a bicycle has wheels of radius 28 cm, take π as $\frac{22}{7}$ to calculate
- a the circumference of a wheel
- b the number of times the wheels rotate when the bicycle travels a distance of 88 metres.

28 This semicircular shape has a diameter of 21 cm. Take π as $\frac{22}{7}$ and calculate the length of
- a its curved side
- b its perimeter.

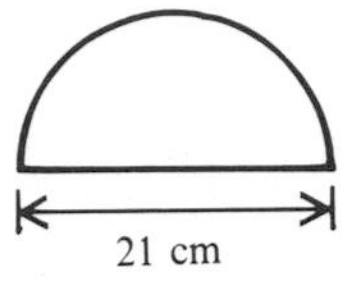

29 Take π as $\frac{22}{7}$ to find the perimeter of a semicircular shape with a diameter of
- a 14 cm
- b 35 cm
- c 42 cm
- d 70 cm.

30 Take π as 3.14 to find the perimeter of a semicircular shape with a diameter of
- a 5 cm
- b 4 cm
- c 8 cm
- d 12 cm.

31 A semicircular protractor has a radius of 4.2 cm. Calculate the length of its perimeter.

32

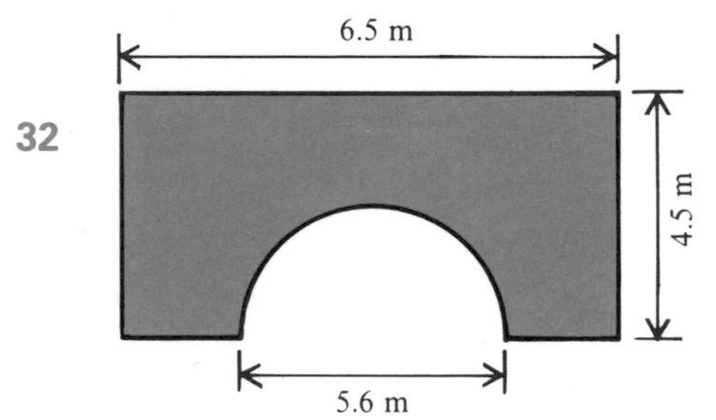

A rectangular sheet of metal (6.5 metres by 4.5 metres) has a semicircular piece of diameter 5.6 m stamped out of it. Find the perimeter of the remaining piece.

33 A belt is connected between two wheels of equal size (radius 38 cm) so that the two straight lengths of belt measure 1.65 metres. Find the total length of the belt.

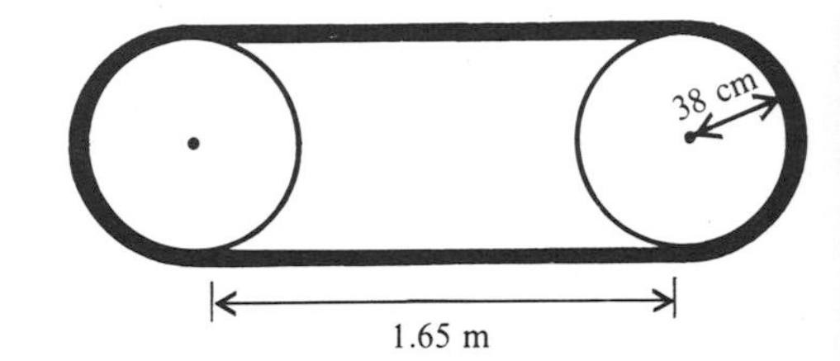

34 The two semicircles in this diagram have diameters AB = 12 cm and CD = 18 cm. If AC = BD = 16 cm, find the perimeter of the shape.

35 In the figure alongside, LM = 24 cm and MN = 12 cm. Calculate
- a the radius of each of the three semicircles
- b the perimeter of the shape.

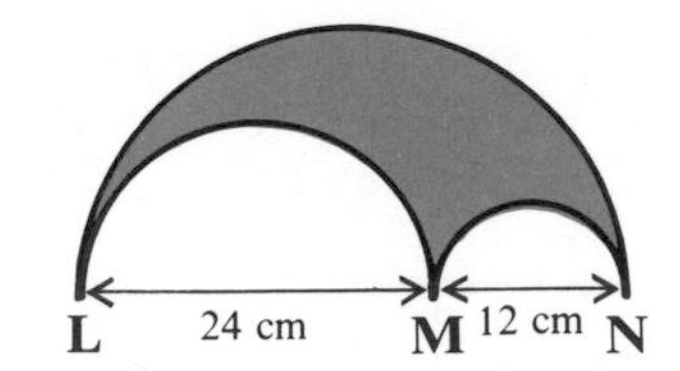

Circumference of circles

36 In the two shapes below, the semicircles have diameters of 18.6 cm and 12.6 cm. Find

a the distance XY b the perimeter of each shape.

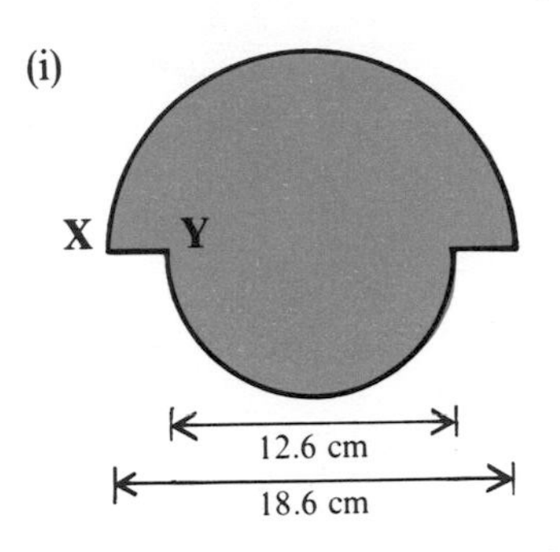

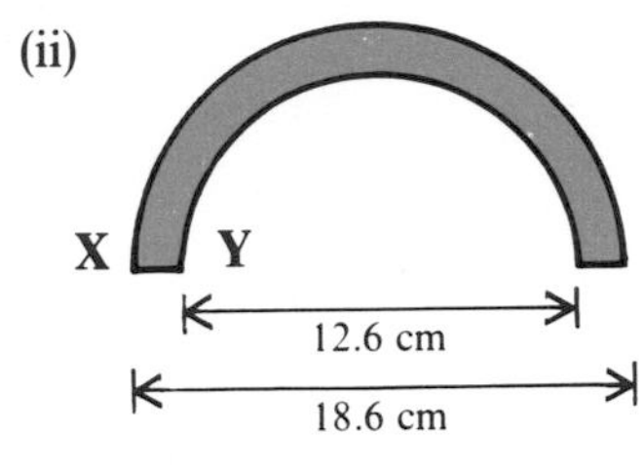

37 This shape is formed from three semicircles whose diameters of length 9 cm make the sides of an equilateral triangle.
Calculate the perimeter of the shape.

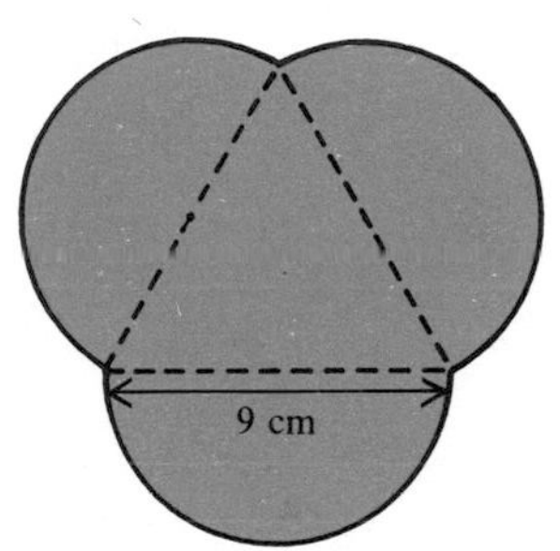

38 Two identical arcs AB and YZ of quadrants of circles of radius 5 cm are joined by two straight lines AY and BZ each 8 cm long.
Find the perimeter of the shape which is made.

39 The coloured shape is formed by removing one semicircle and two identical quadrants from a square of side 15 cm.
Calculate the perimeter of the coloured shape.

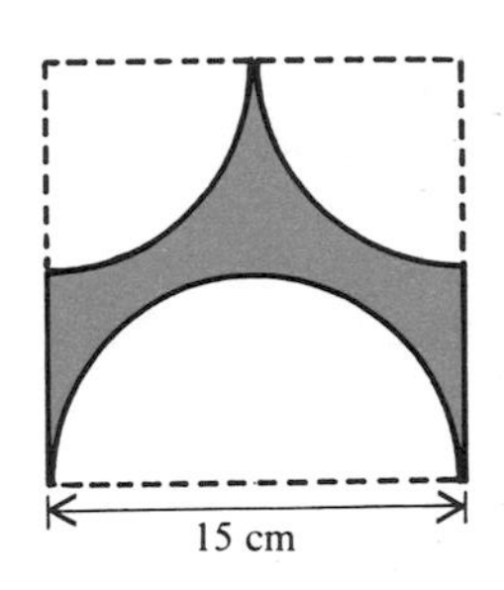

40 This petal can be drawn as one curve without taking pencil from paper by starting from point A and drawing four identical semicircles of diameter 8 cm.
What is the total length of the curve which is drawn?

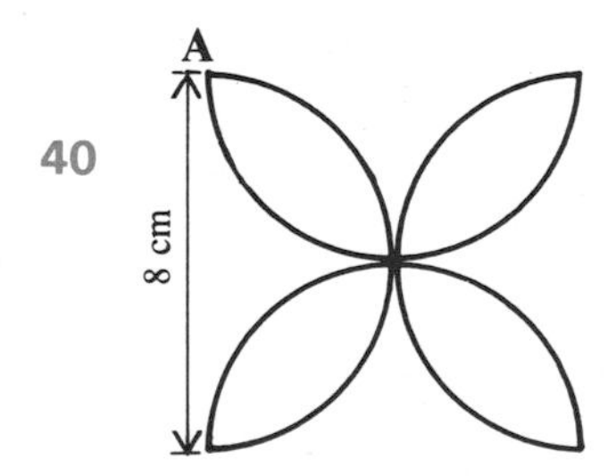

Lengths of arcs

Part 1

1 Find (i) the fraction each arc is of the circumference of the whole circle
(ii) the length of the arc. Take π as $\frac{22}{7}$.

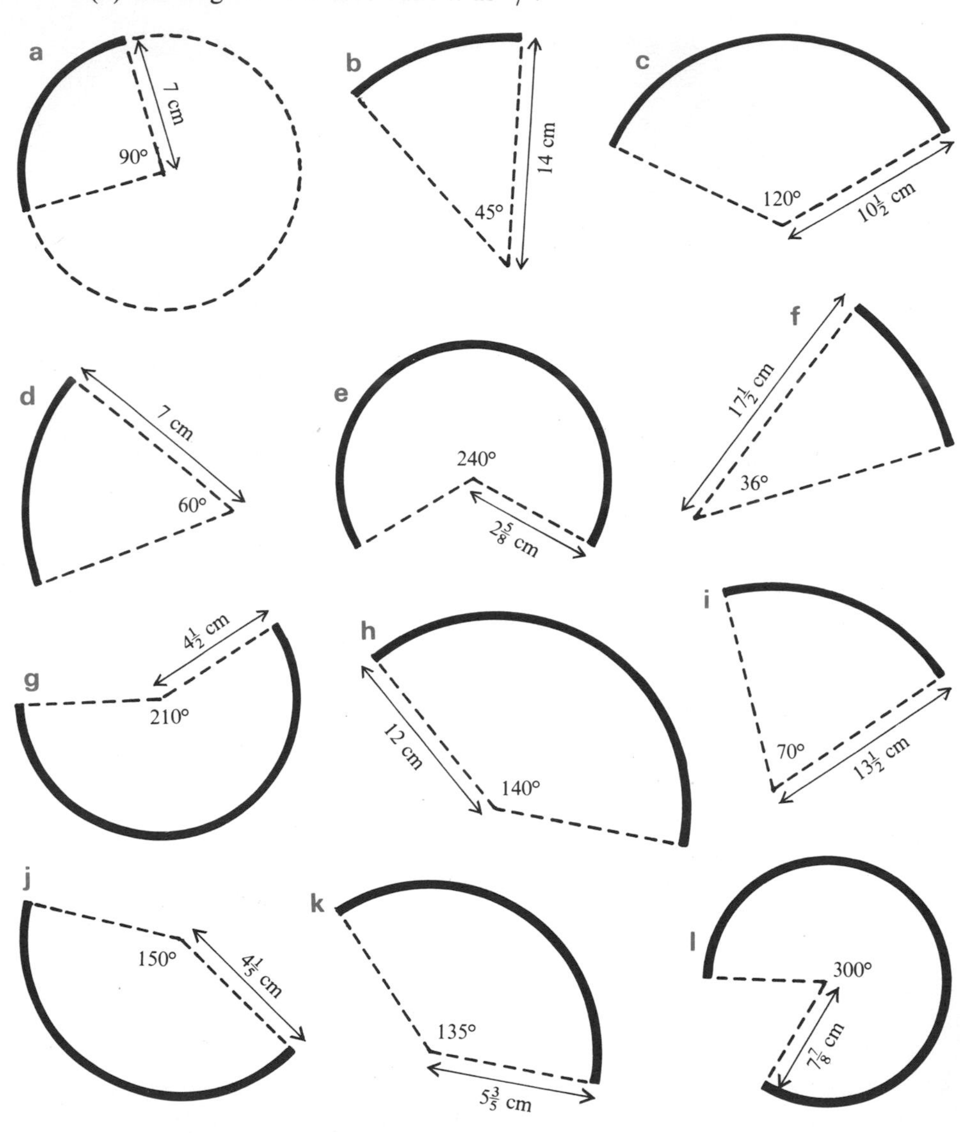

2 Take π as 3.14 to find the length of the arc AB when the radius of the circle and the angle at the centre are respectively

- **a** 2 cm and 60°
- **b** 5 cm and 90°
- **c** 4 cm and 45°
- **d** 10 cm and 120°
- **e** 3 cm and 36°
- **f** 2.5 cm and 240°
- **g** 5.5 cm and 72°
- **h** 1.0 cm and 270°
- **i** 4.5 cm and 135°
- **j** 3.5 cm and 225°.

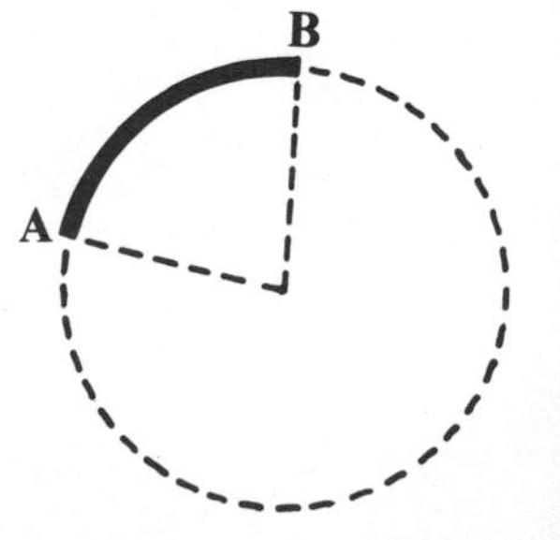

Lengths of arcs

3 Use logarithms or a calculator to find the length of the arc YZ when the radius r and the angle θ at the centre are

a $r = 6.25$ cm and $\theta = 53°$
b $r = 8.12$ cm and $\theta = 85°$
c $r = 7.75$ cm and $\theta = 66°$
d $r = 24.5$ cm and $\theta = 95°$
e $r = 43.7$ cm and $\theta = 23°$ f $r = 39.2$ cm and $\theta = 146°$
g $r = 67.8$ cm and $\theta = 185°$ h $r = 117$ cm and $\theta = 206°$
i $r = 145$ cm and $\theta = 242°$ j $r = 285$ cm and $\theta = 22.5°$
k $r = 72.4$ cm and $\theta = 62.5°$ l $r = 6.24$ cm and $\theta = 71.8°$.

4 Take π as 3.14. Calculate (to 3 significant figures) the length of an arc of a circle of diameter

a 8 cm with an angle of 45° at the centre
b 10 cm with an angle of 60° at the centre
c 24 cm with an angle of 240° at the centre
d 9 cm with an angle of 30° at the centre
e 21 cm with an angle of 270° at the centre
f 15 cm with an angle of 72° at the centre.

5 Use logarithms or a calculator to find the length of an arc of a circle of diameter d with an angle at the centre θ, when

a $d = 6.75$ cm, $\theta = 82°$ b $d = 9.05$ cm, $\theta = 77°$
c $d = 5.94$ cm, $\theta = 94°$ d $d = 27.2$ cm, $\theta = 56°$
e $d = 14.5$ cm, $\theta = 132°$ f $d = 36.4$ cm, $\theta = 146°$
g $d = 20.6$ cm, $\theta = 212°$ h $d = 17.8$ cm, $\theta = 243°$.

6 Find the length of the arc PQ and hence the perimeter of the sector OPQ when

a $r = 8$ cm and $\alpha = 60°$
b $r = 20$ cm and $\alpha = 45°$
c $r = 17.8$ cm and $\alpha = 69°$
d $r = 8.65$ cm and $\alpha = 37°$.

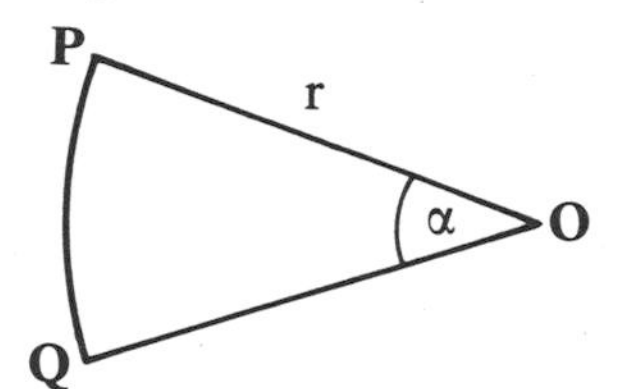

7

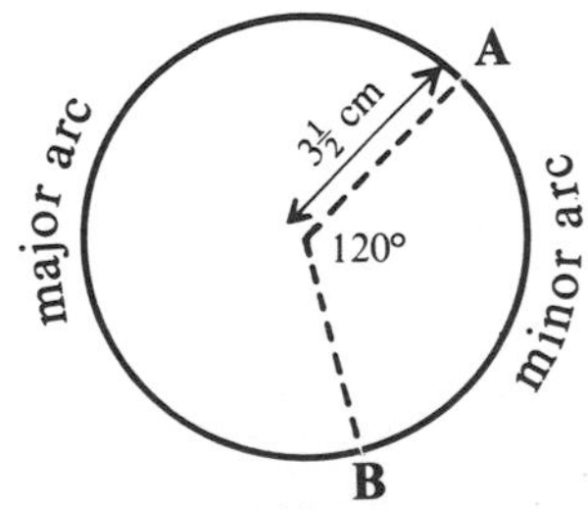

Find the length of the minor arc AB and the major arc AB when the radius of the circle is $3\frac{1}{2}$ cm and the angle at the centre of the minor arc is 120°. Take π as $\frac{22}{7}$.

8 Find the length of the minor and the major arcs of a circle which has a radius of 10 cm, if the angle at the centre of the major arc is 300°.

Take π as 3.14 and give your answers to 3 significant figures.

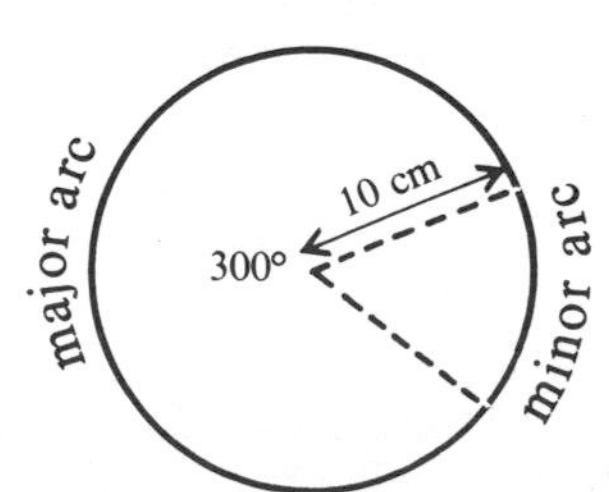

Lengths of arcs

Part 2

Take π as 3.14 unless the problem requests $\frac{22}{7}$.

Give answers to 3 significant figures where appropriate.

1 A road goes round a bend and the white line painted in the middle of the road forms the arc of a circle of radius 120 metres with an angle of 150° at the centre. Calculate the length of the white line on the bend.

2 The pendulum on a clock is 1.2 metres long and its bob swings through an arc of angle 60° as shown. Calculate the distance moved by the bob in one complete swing to and fro.

3 A clock has a minute-hand 12.5 cm long. How far does the tip of it move in 20 minutes?

4 A clock has an hour-hand 4 cm long. Take π as $\frac{22}{7}$ and find how far the tip of the hand moves between 2 pm and 9 pm in the same day.

5

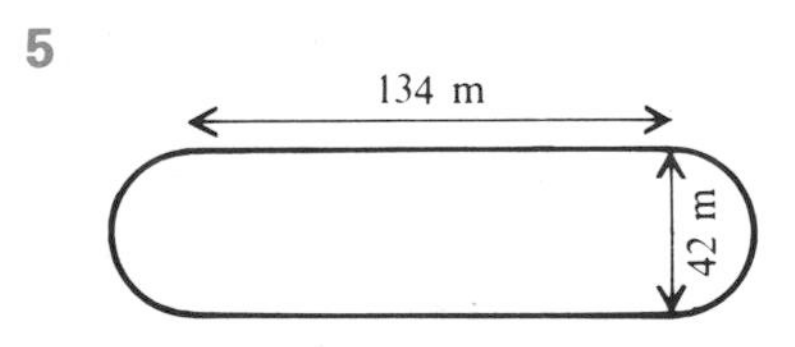

A race-track has two straight sections of 134 metres each and two semicircular ends of diameter 42 metres. Take π as $\frac{22}{7}$ and calculate

a the length of each semicircular end

b the distance covered in one lap of the track

c how many laps are needed in a 2-km race.

6 The diameter of a semicircular protractor is 15 cm. Find the length of the perimeter of the protractor.

7 A circle of radius 9.5 cm has a quadrant cut away. Find the perimeter of the remaining piece.

8

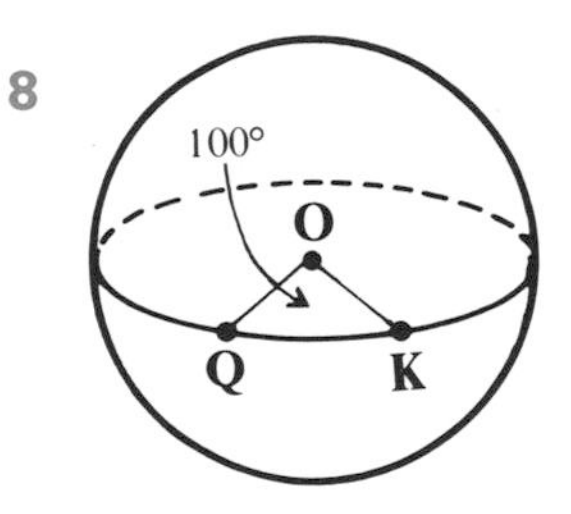

Quito (Q) in Ecuador and Kampala (K) in Uganda both lie on the Equator. Angle QOK is 100° where O is the centre of the earth.

Taking the radius of the earth as 6370 km, calculate the distance between these two cities.

9 London (L) and Accra (A) in Ghana are both on the Greenwich meridian such that angle LOA is 45° where O is the centre of the earth.

If the earth's radius is 6370 km, find the distance between London and Accra.

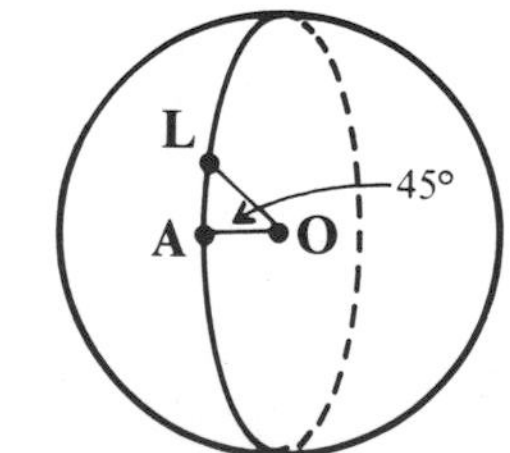

10

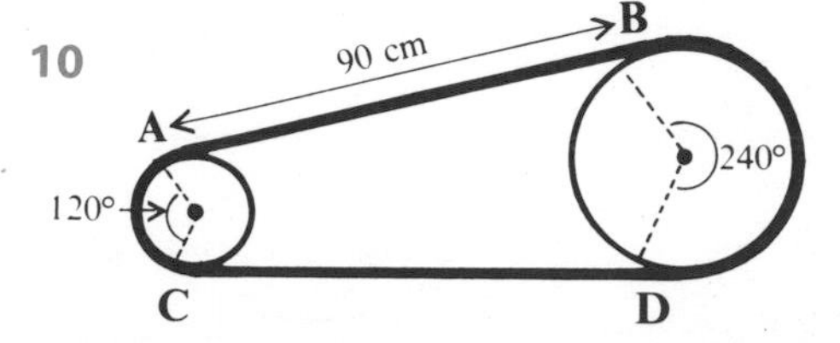

A small driving wheel of radius 40 cm is connected by a belt to a larger wheel of radius 85 cm so that the straight sections of the belt AB and CD are 90 cm long. Find

a the length of the arc AC

b the total length of the belt.

Lengths of arcs

11 The trapdoor ABCD is shown fully open and held in place by the string BB′ such that angle BAB′ is 140°. The door is now closed so that it covers the hole AB′C′D. If AB = 45 cm, take π as $\frac{22}{7}$ to calculate the distance travelled by B as the door closes.

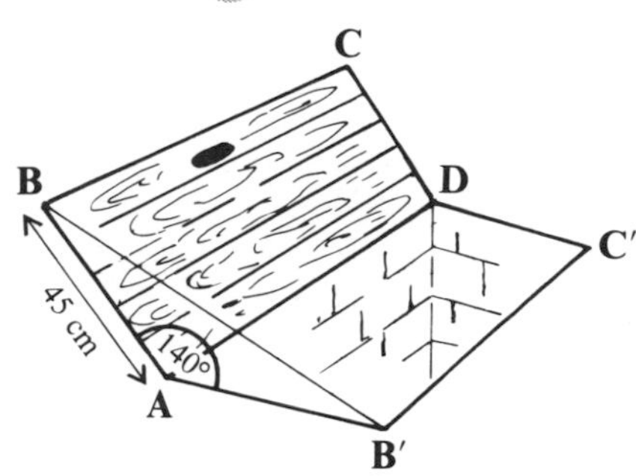

12 A piece of string PQ is fixed at its end P to a cylinder of radius 8 cm and centre O as shown. If Q moves so that the string is wrapped around the cylinder, then Q comes in contact with the cylinder at Q′ where angle POQ′ is 240°. Calculate the length of the string to 3 significant figures.

13 A path is laid in an ornamental garden of the shape shown; that is, two circular arcs of equal circles of radius 16 metres.
Given the angle of 270° as indicated, calculate the length of the path.

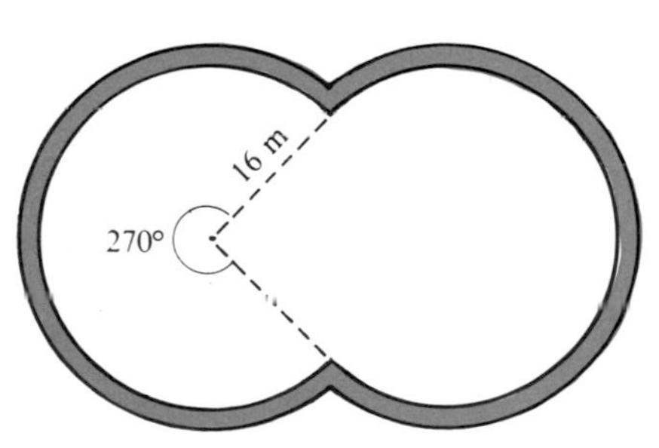

14 A flower-bed is cut in the shape of two squares of side 5 m and two quadrants of a circle. Find

a the length of the arc LM
b the perimeter of the flower bed
c the cost of plastic edging at 24 pence per metre.

15 A shape is stamped out of a sheet of metal. Its dimensions are shown in this diagram. Take π as $\frac{22}{7}$ to calculate the perimeter of the shape.

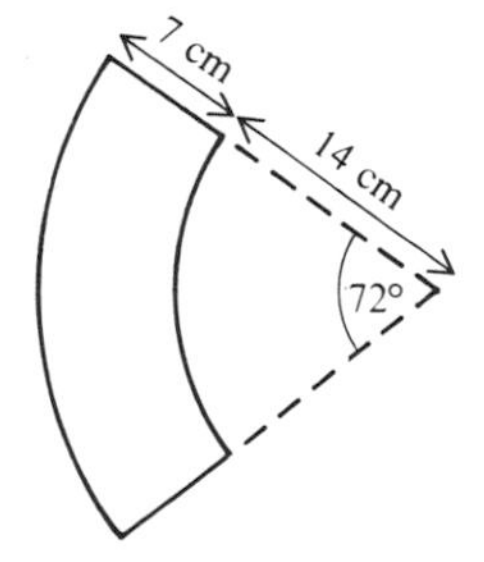

16 Three pencils of radius $\frac{7}{8}$ cm are held together by a rubber band so that their cross-sections are as in this diagram.

a State the size of angle AOB.
b Take π as $\frac{22}{7}$ to find the length of the arc AB.
c State the distance BC.
d How long is the rubber band?

Logarithms

	0	1	2	3	4	5	6	7	8	9
1.0	0.000	004	009	013	017	021	025	029	033	037
1.1	0.041	045	049	053	057	061	064	068	072	076
1.2	0.079	083	086	090	093	097	100	104	107	111
1.3	0.114	117	121	124	127	130	134	137	140	143
1.4	0.146	149	152	155	158	161	164	167	170	173
1.5	0.176	179	182	185	188	190	193	196	199	201
1.6	0.204	207	210	212	215	217	220	223	225	228
1.7	0.230	233	236	238	241	243	246	248	250	253
1.8	0.255	258	260	262	265	267	270	272	274	276
1.9	0.279	281	283	286	288	290	292	294	297	299
2.0	0.301	303	305	307	310	312	314	316	318	320
2.1	0.322	324	326	328	330	332	334	336	338	340
2.2	0.342	344	346	348	350	352	354	356	358	360
2.3	0.362	364	365	367	369	371	373	375	377	378
2.4	0.380	382	384	386	387	389	391	393	394	396
2.5	0.398	400	401	403	405	407	408	410	412	413
2.6	0.415	417	418	420	422	423	425	427	428	430
2.7	0.431	433	435	436	438	439	441	442	444	446
2.8	0.447	449	450	452	453	455	456	458	459	461
2.9	0.462	464	465	467	468	470	471	473	474	476
3.0	0.477	479	480	481	483	484	486	487	489	490
3.1	0.491	493	494	496	497	498	500	501	502	504
3.2	0.505	507	508	509	511	512	513	515	516	517
3.3	0.519	520	521	522	524	525	526	528	529	530
3.4	0.531	533	534	535	537	538	539	540	542	543
3.5	0.544	545	547	548	549	550	551	553	554	555
3.6	0.556	558	559	560	561	562	563	565	566	567
3.7	0.568	569	571	572	573	574	575	576	577	579
3.8	0.580	581	582	583	584	585	587	588	589	590
3.9	0.591	592	593	594	595	597	598	599	600	601
4.0	0.602	603	604	605	606	607	609	610	611	612
4.1	0.613	614	615	616	617	618	619	620	621	622
4.2	0.623	624	625	626	627	628	629	630	631	632
4.3	0.633	634	635	636	637	638	639	640	641	642
4.4	0.643	644	645	646	647	648	649	650	651	652
4.5	0.653	654	655	656	657	658	659	660	661	662
4.6	0.663	664	665	666	667	667	668	669	670	671
4.7	0.672	673	674	675	676	677	678	679	679	680
4.8	0.681	682	683	684	685	686	687	688	688	689
4.9	0.690	691	692	693	694	695	695	696	697	698
5.0	0.699	700	701	702	702	703	704	705	706	707
5.1	0.708	708	709	710	711	712	713	713	714	715
5.2	0.716	717	718	719	719	720	721	722	723	723
5.3	0.724	725	726	727	728	728	729	730	731	732
5.4	0.732	733	734	735	736	736	737	738	739	740

Logarithms

	0	1	2	3	4	5	6	7	8	9
5.5	0.740	741	742	743	744	744	745	746	747	747
5.6	0.748	749	750	751	751	752	753	754	754	755
5.7	0.756	757	757	758	759	760	760	761	762	763
5.8	0.763	764	765	766	766	767	768	769	769	770
5.9	0.771	772	772	773	774	775	775	776	777	777
6.0	0.778	779	780	780	781	782	782	783	784	785
6.1	0.785	786	787	787	788	789	790	790	791	792
6.2	0.792	793	794	794	795	796	797	797	798	799
6.3	0.799	800	801	801	802	803	803	804	805	806
6.4	0.806	807	808	808	809	810	810	811	812	812
6.5	0.813	814	814	815	816	816	817	818	818	819
6.6	0.820	820	821	822	822	823	823	824	825	825
6.7	0.826	827	827	828	829	829	830	831	831	832
6.8	0.833	833	834	834	835	836	836	837	838	838
6.9	0.839	839	840	841	841	842	843	843	844	844
7.0	0.845	846	846	847	848	848	849	849	850	851
7.1	0.851	852	852	853	854	854	855	856	856	857
7.2	0.857	858	859	859	860	860	861	862	862	863
7.3	0.863	864	865	865	866	866	867	867	868	869
7.4	0.869	870	870	871	872	872	873	873	874	874
7.5	0.875	876	876	877	877	878	879	879	880	880
7.6	0.881	881	882	883	883	884	884	885	885	886
7.7	0.886	887	888	888	889	889	890	890	891	892
7.8	0.892	893	893	894	894	895	895	896	897	897
7.9	0.898	898	899	899	900	900	901	901	902	903
8.0	0.903	904	904	905	905	906	906	907	907	908
8.1	0.908	909	910	910	911	911	912	912	913	913
8.2	0.914	914	915	915	916	916	917	918	918	919
8.3	0.919	920	920	921	921	922	922	923	923	924
8.4	0.924	925	925	926	926	927	927	928	928	929
8.5	0.929	930	930	931	931	932	932	933	933	934
8.6	0.934	935	936	936	937	937	938	938	939	939
8.7	0.940	940	941	941	942	942	943	943	943	944
8.8	0.944	945	945	946	946	947	947	948	948	949
8.9	0.949	950	950	951	951	952	952	953	953	954
9.0	0.954	955	955	956	956	957	957	958	958	959
9.1	0.959	960	960	960	961	961	962	962	963	963
9.2	0.964	964	965	965	966	966	967	967	968	968
9.3	0.968	969	969	970	970	971	971	972	972	973
9.4	0.973	974	974	975	975	975	976	976	977	977
9.5	0.978	978	979	979	980	980	980	981	981	982
9.6	0.982	983	983	984	984	985	985	985	986	986
9.7	0.987	987	988	988	989	989	989	990	990	991
9.8	0.991	992	992	993	993	993	994	994	995	995
9.9	0.996	996	997	997	997	998	998	999	999	1.000
10.0	1.000									

Conversions and Constants

Length

10 mm = 1 cm
100 cm = 1 m
1000 m = 1 km

Area

100 mm^2 = 1 cm^2
10 000 cm^2 = 1 m^2
10 000 m^2 = 1 hectare (ha)
100 ha = 1 km^2

Volume/capacity

1000 cm^3 = 1 litre
1000 litres = 1 m^3

Mass

1000 grams = 1 kg
1000 kg = 1 tonne

Imperial–metric conversions

Length	1 inch	= 2.54 cm	1 cm	= 0.394 inches
	1 foot (= 12 inches)	= 30.5 cm		
	1 yard (= 3 feet)	= 0.914 m	1 m	= 1.09 yards
	1 mile (= 1760 yards)	= 1.61 km	1 km	= 0.621 miles
			8 km	= 5 miles
Area	1 sq. inch	= 6.45 cm^2	1 cm^2	= 0.155 sq. inch
	1 sq. foot	= 929 cm^2		
	1 sq. yard	= 0.836 m^2	1 m^2	= 1.20 sq. yards
	1 acre (= 4840 yd^2)	= 0.405 ha	1 ha	= 2.47 acres
	1 sq. mile (= 640 acres)	= 2.59 km^2	1 km^2	= 0.386 sq. miles
Volume/capacity				
	1 cubic inch	= 16.4 cm^3	1 cm^3	= 0.061 cu. inches
	1 cubic yard	= 0.765 m^3	1 m^3	= 1.31 cu. yards
	1 pint	= 0.568 litres	1 litre	= 1.76 pints
	1 gallon (= 8 pints)	= 4.55 litres	1 litre	= 0.22 gallons
Mass	1 ounce	= 28.35 grams	1 gram	= 0.0353 ounces
	1 pound (= 16 ounces)	= 0.454 kg	1 kg	= 2.20 pounds
	1 ton (= 2240 pounds)	= 1.02 tonnes	1 tonne	= 0.984 tons

The metric prefixes

deca	da	10	deci	d	10^{-1}
hecto	h	10^2	centi	c	10^{-2}
kilo	k	10^3	milli	m	10^{-3}
mega	M	10^6	micro	μ	10^{-6}
giga	G	10^9	nano	n	10^{-9}
tera	T	10^{12}	pico	p	10^{-12}

Constants (to 3 significant figures)

π = 3.14
$\log \pi$ = 0.497
1 radian = 57.3°
Radius of the Earth = 6370 km
Acceleration due to gravity = 9.81 m/s^2
Velocity of light = 3×10^8 m/s